Longman Geography
for IGCSE

Authors
John Pallister
Ann Bowen
Roger Clay
Carmela Di Landro
Olly Phillipson
Steve Milner

PEARSON
Longman

Edinburgh Gate
Harlow, Essex

Contents

PEARSON EDUCATION LIMITED

Edinburgh Gate, Harlow, Essex, CM20 2JE, and associated companies throughout the world.

© Pearson Education Limited 2006

ISBN 978-1-4058-0209-3

Fourth impression 2009

Printed in China
SWTC/04

The Publishers' policy is to use paper manufactured from sustainable forests.

Produced for publication by Start to Finish

Illustrations by Judy Brown, Tom Cross, Hardlines, Nick Hawken, Moondisks Ltd, Pat Murray/Graham Cameron Illustration, Oxford Illustrators

We are grateful to the following for permission to reproduce photographs and other copyright material.

(t) = top, (b) = bottom, (m) = middle, (l) = left, (r) = right

ACESTOCK.COM: pg72; **Adams Picture Library:** pg155(b) (Mark Follon); **"Afan Digital"/ Photographersdirect.com:** pg113(b); **Alamy:** pgs14(t) (Robert Preston), 14(b) (David Martyn Hughes), 67 (©David Hoffman), 173(t) (G P Bowater), 177(t) (Peter Casolino), 191(b) (©Ken Welsh); **Art Directors & TRIP:** pgs29(t) (B Ashe), 46(b) (Tibor Bognar), 155(t) (S Samuels); **Big Pit, Blaenarvon, Wales:** pgs84(t) & 112; **©Birdlife South Africa:** pg 181; **www.JohnBirdsall.co.uk:** pg149; **BrazilPhotoBank:** pg143(b); **Camera Press London:** pgs40 & 44; **Roger Clay:** pgs63(l) & 193(l); **John Cleare Mountain Camera:** pg11; **Corbis:** pgs22 (©Larry Lee Photography), 103 (©Reuters), 128 (©Fred Greaves), 141(t) (©Sergio Dorantes), 142 (©Paulo Fridman),148 (©Morton Beebe), 180(t) (©Jagadeesh/Reuters), 186 (©Michael S Yamashita); **Sylvia Cordaiy Photo Library:** pg19 (Geoffrey Taunton); **Sue Cunningham Photographic:** pgs166(t) & 199(t) (SCP); **Department for International Development © Crown Copyright:** pg198(l & r); **Ecoscene:** pgs88 (Paul Thompson); 166(b) (E J Bent); **© EMPICS:** pgs46(t), 52, 53(t), 117, 131, 180(b); **Eye Ubiquitous:** pgs57(l) (Hutchison), 63(r) (©Gerald Fritz), 137(©Edward Parker); **Fairtrade Foundation:** pg197; **Fundu Lagoon, Pemba Island:** pg193(r) (photo Ken Niven); **Getty Images:** pgs5, 18, 53(b), 83, 135; **Paul Glendell/Photographersdirect.com:** pg23; **Reproduced by kind permission of GMPTE:** pg156(b); **Green and Black's:** pg61(b); **Richard Greenhill:** pg77(b); **Robert Harding Picture Library:** pg85(b) (Leslie Evans); **Imagestate:** pg141(b); **Intermediate Technology:** pg70; **International Rice Research Institute, Philippines:** pg69; **Magnum:** pg97(l) (E Reed); **Aidan O'Rourke www.aidan.co.uk:** pg156(t); **Oxford Scientific:** pg166(m); **John Pallister:** pgs6, 32, 77(t), 78(l & r), 87(t), 89, 121(t), 144(m & b), 146, 153, 185; **Panos:** pgs25(b) & 140(b) (Mark Henley), 97(r) (Sean Sprague), 163 (Jeremy Hartley); **Rex Features:** pgs 15, 24(r), 57(r), 74, 111, 125, 179; **Reuters:** pg45; **Science Photo Library:** pgs37 (CNES, 1989 Distribution Spot Image), 38 (David Parker), 49 (NOAA), 85(t) (Maximilian Stock Ltd); **©Skyscan:** pgs24(l) (K Allen), 64 (l & r) (Skyviews Aerial Archive), 75, 86(b) (Patrick Roach), 113(t), 150 (K Hallam), 157(t) (W Cross); **South American Pictures:** pgs138 & 173(b) (Tony Morrison); **©Still Pictures:** pgs51(inset), 65 (Thomas Raupach), 73 (Ron Giling/Lineair), 85(m), 87(b), 91 & 169 (t & b) (All Mark Edwards), 86(t) (Edward Parker), 87(ml) & 90 (David Brain), 121(b) (George Mulala/Lineair), 159 (Martin Bond), 171 (Nigel Dickinson), 172 (Alan Watson), 175 (Kim Heacox), 177b (S J Krasemann), 189 (Thomas Bauer), 191(t) (Hartmut Schwarzbach/Argus), 195 (Lineair), 199(b) (Jorgen Schytte), 203(t) (Wolfgang Schmidt); **Topfoto:** pgs25(t), 101, 129(b) & 143(t) (All Imageworks), 47(t & bl), 61(t) & 130 (AP), 93, 151 (PA), 157(b) (Lowe); **Travel Ink:** pg105 (Stephen Psallidas), 123 (Dorothy Burrows); **USDD:** pg51; **©World Bank:** pg200.

Picture Research by: Sandie Huskinson-Rolfe of **PHOTOSEEKERS**.

Front cover image © John Walmsley, Education Photos

Every effort has been made to trace the copyright holders, and we apologise in advance for any unintentional omissions. We would be pleased to insert the appropriate acknowledgement in any subsequent edition of this publication.

Water

Unit Contents

The Victoria Falls on the Zambezi River, Africa. Where and why do rivers form spectacular waterfalls?

1.1 The hydrological cycle and drainage basins

Fresh water is essential for life on earth. Water that reaches the land surface forms part of the **water cycle**, also called the **hydrological cycle** (Source 1). The main input into the system is **precipitation**. This is usually rain, but in high mountain areas snow is frequent as well. Water may flow quickly through the system as **runoff** on the surface. Energy from the sun **evaporates** sea water: it changes liquid water into water vapour in the atmosphere. As the water vapour is drawn higher up into the atmosphere, it is cooled. The water vapour may **condense** into water droplets which can be seen as clouds. Precipitation falls from clouds that are sufficiently tall and thick, and the water cycle begins all over again.

The water cycle is more complicated than this, however. Some of the rain water may never reach the sea; instead it is lost directly back into the atmosphere from the leaves of plants. This process is known as **evapo-transpiration**. Some of the rain water is **intercepted** by trees so that its flow is delayed. Precipitation that falls as snow can be stored in glaciers on the surface (Source 2). Rainwater may be stored in lakes. Some rain water seeps down through the soil. This is the process of **infiltration**. Some of the infiltrated water seeps further down to fill empty spaces in the rock, which is known as **percolation**. The water can only do this until it reaches the level called the **water table**. Below this level the spaces in the rock have already been filled with water. At this point the water flows sideways as **groundwater flow**, as Source 1 shows.

Source 2 | Perito Moreno glacier

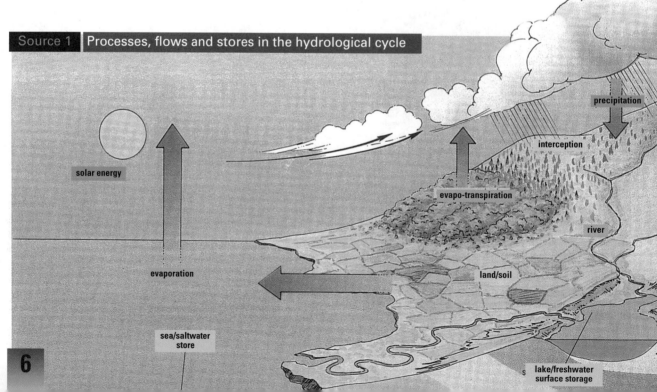

Source 1 | Processes, flows and stores in the hydrological cycle

precipitation

interception

solar energy

evapo-transpiration

river

evaporation

land/soil

sea/saltwater store

lake/freshwater surface storage

Source 3 | The drainage basin

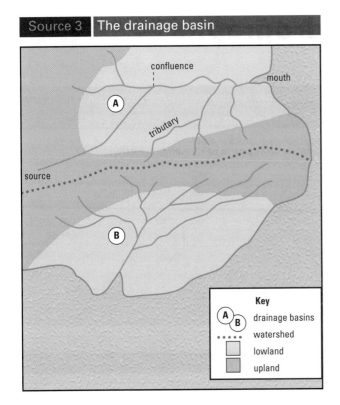

Key

A B — drainage basins

• • • • — watershed

☐ lowland

☐ upland

Source 4 | Factors affecting the rate of runoff

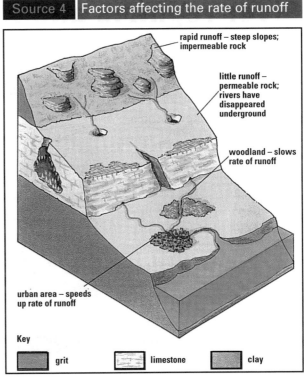

rapid runoff – steep slopes; impermeable rock

little runoff – permeable rock; rivers have disappeared underground

woodland – slows rate of runoff

urban area – speeds up rate of runoff

Key

☐ grit   ☐ limestone   ☐ clay

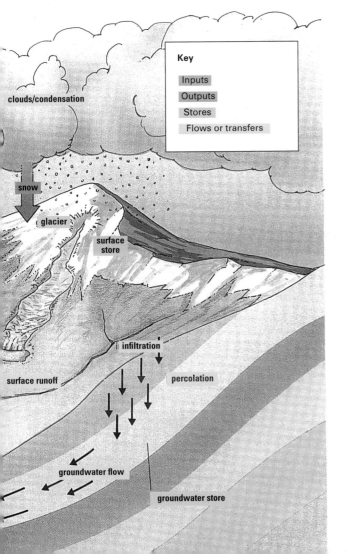

Key

Inputs

Outputs

Stores

Flows or transfers

clouds/condensation

snow

glacier

surface store

surface runoff

infiltration

percolation

groundwater flow

groundwater store

Each river has its own **drainage basin**. Each drainage basin has its own inputs, flows, stores and outputs. As Source 3 shows, it is possible to draw a definite dividing line between each drainage basin. It follows the tops of the hills and is called the **watershed**. The main river has its source in the higher parts of the basin where most precipitation falls. Smaller streams or **tributaries** join up with the main river; they meet at a **confluence**. The **mouth** of the river is where it meets the sea.

Each drainage basin has its own features of rock type, relief (shape of the land) and land use and these affect how quickly or how slowly the water moves through the basin. Source 4 shows how the features of a drainage basin can affect runoff. The rock type and relief are physical factors over which humans have little influence. However, land use is different. Land uses can be changed by people. Urban areas increase the rate of runoff. The rain water hits solid surfaces such as roofs, pavements and roads; the water is led into drains which speed up its overland flow into rivers.

1.2 River regimes and hydrographs

Rivers occupy their own, separate **drainage basins**. The streams and tributaries which make up each river system (Source 3) collect water from inside the drainage basin – an area of land known as the **catch-ment area**. It is important that we know how quickly any water falling in the catchment area will reach the river. If it reaches the river quickly, flooding may also happen quickly if there is too much water for the river channel to hold.

It is important to know how much water a river can hold. This is known as the river's **discharge** and is measured in cumecs (cubic metres per second). It is important to measure and study river flow to help predict where and when flooding may occur. The flow will usually vary from month to month (Source 1). This variation is called the **river regime**. In most rivers, it closely reflects local climatic conditions. For example, in the UK river discharge is generally at its highest in the winter because this is when rainfall is highest, there is less vegetation to intercept rain and colder temperatures result in low evapotranspiration.

Average monthly discharge for the River Ganges, India and Bangladesh

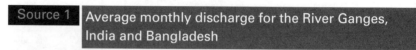
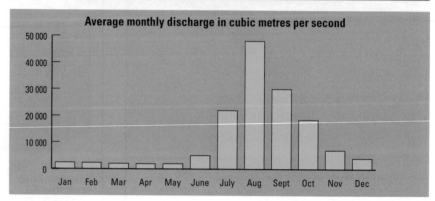

Average monthly discharge in cubic metres per second

Source 2 | Storm hydrograph

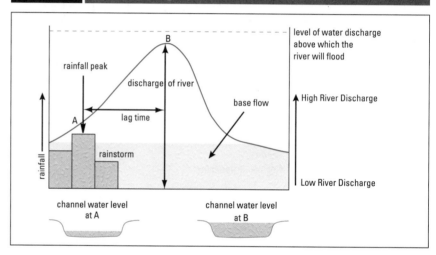

Source 3 | Storm hydrograph for a river in flood

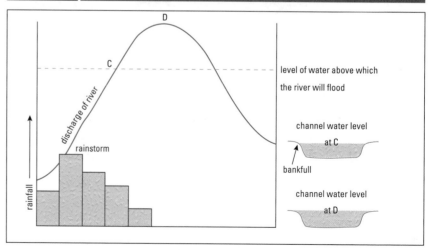

Hydrographs

Storm hydrographs (Source 2) plot the increase in a river's discharge following a storm. It will take some time after the storm for river levels to rise, this is called the **lag time**. The lag time is the time between peak rainfall and peak discharge. The shorter the lag time, the quicker the water reaches the river channel giving a steep curve on the hydrograph (Source 3). This can cause sudden flooding.

Variations in discharge

There are many factors which affect the amount of discharge, the speed at which water reaches the river and the chances of flooding (Source 4). Some are natural, but many are influenced by human activity.

- **Climate:** The amount and intensity of the rain has a major impact on discharge. Heavy rain may not sink into the ground but flow quickly on the surface to rivers. If precipitation is in the form of snow, it could be weeks before it melts. If the ground is frozen, melting snow will reach the river quickly.
- **Vegetation:** Lots of trees and plants will intercept and delay rain reaching the ground. Bare ground will speed up runoff.

- **Relief/slope:** Steep slopes cause rapid surface runoff, so water will reach the river more quickly. Flat land may lead to water infiltrating the soil more slowly and therefore delaying it reaching the river.
- **Geology/rock type:** The types of rock which make up individual river basins have a major influence on discharge. Impermeable rock will not allow water to sink into it, so will speed up runoff, and is more likely to lead to flooding. Permeable rock will allow infiltration and percolation of water in the soil and underground, slowing down its journey to the river channel.
- **Size and density of drainage basin:** In larger drainage basins, water generally takes longer to reach the river than in smaller ones, although drainage density will also influence this – the higher the density, the quicker discharge will rise.
- **Land use:** Areas covered by tarmac and concrete, for example urban areas, speed up run-off when compared to open ground and fields.
- River management schemes like dams will also affect discharge.

In practice, river discharge is affected by a combination of several, if not all, these factors.

Source 4 | A river in flood

The river's course from source to mouth is summarised by its **long profile** (Source 1). The profile is steep and irregular when the river is flowing well above sea level in the uplands, but much gentler and smoother as the river nears the sea.

The main work of any river is to transfer rain water from land to sea. Most of its energy is used simply to keep the water flowing, because it needs to overcome the friction of the river's bed and banks. It may have some spare energy to transport its load of fine material and pebbles. There are four ways in which a river transports its load: solution, suspension, saltation and traction. Source 2 shows how each of these works.

The river is, therefore, an **agent of transport**, carrying material from upland to lowland regions. It is also an **agent of erosion** and **deposition**. Which of these processes is dominant depends mainly on the geology, relief (slope) and velocity of the river and river basin. If rocks are very hard and resistant, there will be less erosion compared to softer rock. In the upper course, the river is running quickly because of steeper gradients, so erosion (Source 3) and transport are important. As it reaches the middle course, the gradient lessens, the river slows down and has less energy to erode – it starts to deposit material it is transporting. In its lower course, close to the sea, the river flows at it slowest, depositing fine grained material.

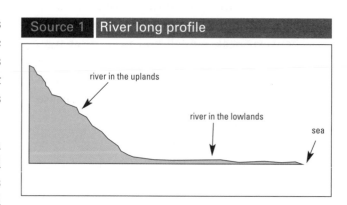

Source 1 | **River long profile**

river in the uplands

river in the lowlands

sea

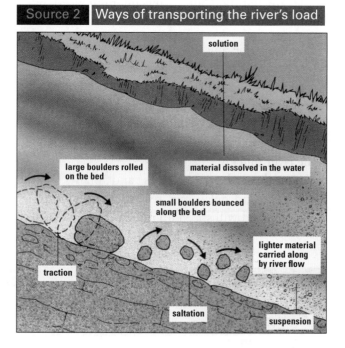

Source 2 | **Ways of transporting the river's load**

solution

large boulders rolled on the bed

material dissolved in the water

small boulders bounced along the bed

lighter material carried along by river flow

traction

saltation

suspension

Source 3 | **Process of river erosion**

Abrasion or corrosion	Hydraulic action
The pebbles being transported remove material from the bed and banks of the river channel, by wearing them away.	The sheer force of the water by itself may be sufficient to dislodge material and erode the bed and sides of the channel.
Solution or corrosion	**Attrition**
Some rocks are subject to chemical attack: chalk and limestone, for example, slowly dissolve in water.	The particles are knocked about as they are being transported. They are reduced in size to sand and eventually to even smaller silt-sized particles. These are more easily moved by the water.

Source 4 shows a river in the uplands. Labels have been added which describe the channel and valley features.

Source 4 Moffat Water in the Southern Uplands

steep-sided, V-shaped valley

waterfall

interlocking spurs

River landforms in the uplands

The main landforms found in the uplands – steep **V-shaped valley**, **interlocking spurs**, **waterfall** and **gorge** (Source 5) – have all been formed by the processes of river erosion already referred to.

The steep-sided V-shaped valley is formed by **vertical erosion**. The river is flowing so high above sea level that its main work is cutting downwards, which is what vertical erosion means. By the processes of abrasion and hydraulic action, the river erodes the rocks on its bed making the valley deeper. There is mass movement of material down the sides of the valley because the valley is so steep and deep.

Interlocking spurs are formed where the river swings from side to side. Again the main work of the river is vertical erosion into the rock on its bed by abrasion and hydraulic action. This means that the river cuts down to flow between spurs of higher land on alternate sides of the valley.

Waterfalls occur where a hard band of rock outcrops which is much more resistant to erosion than the softer rock below it. The river can only slowly erode the hard band of rock; it can erode more quickly the soft rock below. The soft rock is eroded also by the force of the water as it falls, which creates a **plunge pool** at the bottom of the falls. The waterfall gradually retreats upstream leaving a gorge below it. The gorge is protected from erosion by its capping of hard rock.

Source 5 | Formation of valley landforms in upland areas

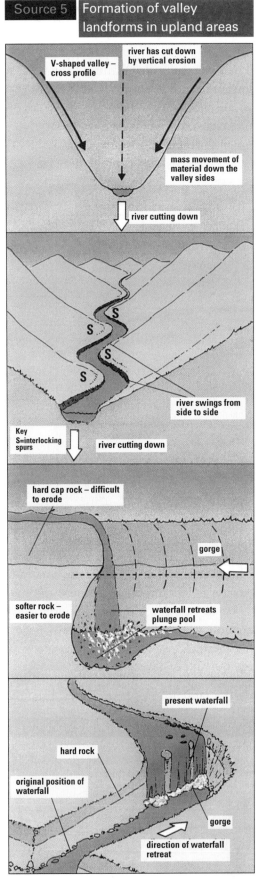

V-shaped valley – cross profile

river has cut down by vertical erosion

mass movement of material down the valley sides

river cutting down

S

river swings from side to side

Key
S=interlocking spurs

river cutting down

hard cap rock – difficult to erode

gorge

softer rock – easier to erode

waterfall retreats plunge pool

present waterfall

hard rock

original position of waterfall

gorge

direction of waterfall retreat

Rivers and river landforms 2: lowlands

The river and valley features (Source 1) change as lowland regions are reached: the river channel is wider and deeper; there are fewer large stones in the bed and river flow can be as fast as in the uplands despite the more gentle gradient; the plan of the river is less straight because of the many **meanders**; the river can often split up into **distributaries** near the sea to form a **delta**; the valley cross-section is wider and flatter, and includes the **floodplain** where the distinctive landforms include **levees** and **ox-bow lakes**.

River and valley features

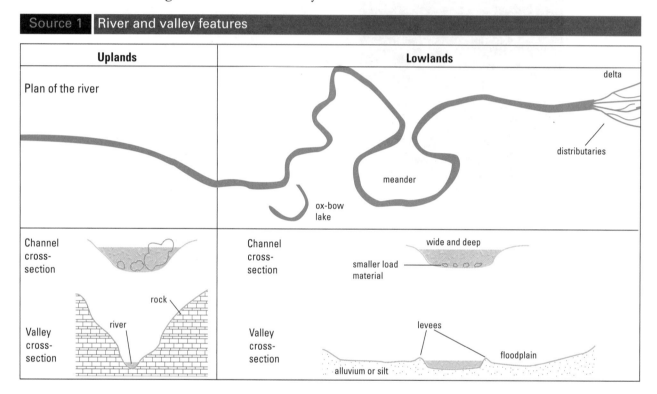

The river is still an agent of erosion, but vertical erosion is less important because the river is too close to sea level. More important is **lateral erosion** where the river wears away the sides of the channel, especially on the outside of bends.

The river becomes an agent of **deposition** as well. Such a large load of material has been picked up

that, once the river loses energy, it drops some of the material it is transporting. Energy is lost when the river flow gets slower, such as on the inside of a bend or where the river meets the sea.

The greatest thickness of river-deposited material, called **alluvium**, is on the floodplain. As its names suggests, the floodplain is an area of flat

Source 2   Formation of the floodplain and levees

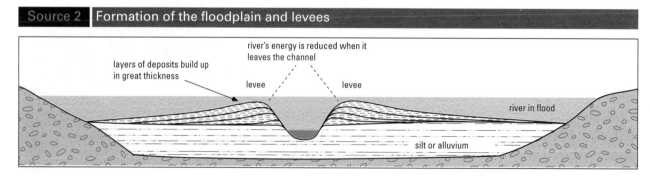

land formed by flooding. Every time the river leaves its channel, it is slowed down and begins to deposit **silt** across the valley floor. A great thickness of alluvial material builds up. The largest amount of deposition is always on the banks of the channel, which builds up to a greater height than the rest of the floodplain to form levees (Source 2).

Meanders and ox-bow lakes

A study of the formation of meanders and ox-bow lakes shows how the river both deposits and erodes laterally (Source 3).

The force of the water undercuts the bank on the outside of a bend to form a steep bank to the channel, called a **river cliff**. An underwater current with a spiral flow carries the eroded material to the inside of the bend where the flow of water is slow. Here the material is deposited to form a gentle bank, called a **slip-off slope**.

The bend of the meander becomes even more pronounced as lateral erosion continues (Source 4). Especially in times of flood, when the river's energy is much greater, the narrow neck of the meander may be broken so that the river flows straight again. This forms an ox-bow lake by cutting off the old meander loop. Deposition during the flooding may seal off the edges of the lake. The lateral erosion on the outside bank of the meander helps to widen the floodplain.

| Source 3 | Formation of meanders and ox-bow lakes |

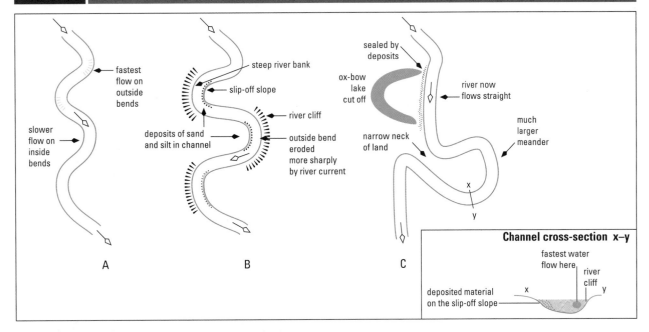

| Source 4 | A meandering river |

Most British rivers are relatively small and go out to sea through estuaries. There are often one or two deep water channels between extensive deposits of sand and mud. An example is the River Tay on pages 16–17. Bigger rivers in other parts of the world may split up just before reaching the sea and form a delta. The River Ganges, which is described on pages 18–19, does this.

Weathering and mass movement

Whilst running water has a major role in the shaping of landforms, once formed the processes of weathering and mass movement can cause further changes. Both processes are heavily influenced by the relief and geology of the area.

Weathering

Weathering is the name given to the break-up and disintegration of rocks on the earth's surface *in situ* (where they are) without movement taking place. Weathered material is frequently transported and eroded further, especially by rivers, waves and glaciers.

There are two main types of weathering:

- mechanical or physical weathering
- chemical weathering.

Mechanical or physical weathering causes rocks to disintegrate, breaking up into smaller fragments. Freeze-thaw, exfoliation and biological weathering are all examples of this type of weathering.

Freeze-thaw takes place when cracks in rocks collect water which then freezes, usually at night when temperatures fall (Source 1). This causes the water to expand and split the rock. As temperatures rise again during the day, the ice melts and pressure is released. This can happen many times until large fragments of rock may break away. These fragments are often found at the bottom of steep mountain slopes and are known as **scree**. This type of weathering is also known as frost shattering.

Hot deserts have a large daily (or diurnal) range of temperatures which cause rocks to disintegrate by **exfoliation** or onion skin weathering (Source 2). Rocks are heated and expand during the day and

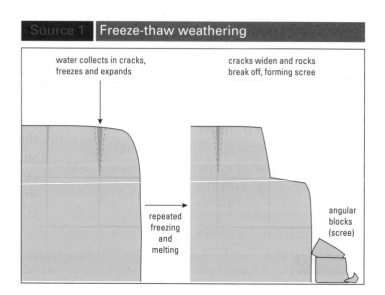

Source 1 | Freeze-thaw weathering

water collects in cracks, freezes and expands

cracks widen and rocks break off, forming scree

repeated freezing and melting

angular blocks (scree)

Source 2 | Rock exfoliation, Namibia

Source 3 | Limestone pavement, caused by rainwater widening joints in carboniferous limestone

cool down at night. This causes stresses in the rock resulting in the peeling away of the outer layer, producing rounded landforms. To take effect, small amounts of water are needed. Trees and shrubs can also cause rocks to disintegrate (biological weathering) as their roots grow down through cracks, widening them and causing them to split and break away.

Chemical weathering involves changes in the actual composition of rocks, usually involving contact with water, so the effect depends on the type of rock being weathered. One of the most common forms of chemical weathering is **solution**. This occurs when rocks like limestone and chalk, which contain calcium carbonate ($CaCO_3$), react with rain. A weak form of carbonic acid is found in rainwater and this will gradually dissolve the rock. Weathered particles are removed by running water, causing further changes to the rock, helping create the distinctive landforms or karst scenery found in carboniferous limestone regions (Source 3).

Mass movement

Once material has been weathered and broken off from rock surfaces, it may start to move downhill. This movement is happening continuously down slopes through gravity, so the steeper the slope the faster the movement will be. The speed at which this happens varies from the relatively slow process of **soil creep**, where soil gradually moves downhill forming small terraces, to potentially dangerous **mudslides** (Source 4) and **landslides**. These are all examples of mass movement.

Many of the deaths caused by Hurricane Mitch in Central America in 1998 (see pages 50–1) were the result of mudslides caused by torrential rain. The rain saturated the ground causing material to break off and slide rapidly downhill. Landslides are sudden movements of masses of material resulting from earth movements like earthquakes or tremors.

A river and its valley
the River Tay, UK

In the uplands

The River Tay is fed by streams which drain the slopes of the Grampian mountains in the Highlands of Scotland. Precipitation in the upland parts of the drainage basin is high (well over 1000 mm per year) and slopes are steep, which gives high amounts of runoff. It is already a big river, about 100 metres across in that part of its course shown on the Ordnance Survey map extract (Source 1).

The valley cross-section is shown in Source 2a. It is V-shaped and steep-sided. The river fills the valley floor. The cross-section shows that the river is still flowing at some height above sea level; it is cutting down into the valley floor by vertical erosion.

As for the land uses shown on the map (Source 1), the small settlements and roads are concentrated on the less steep and more sheltered land in the Tay valley.

To the north of the river much of the land is likely to be used for nothing better than rough grazing for sheep and deer. South of the river coniferous (evergreen) woodland is the land use which covers the largest area.

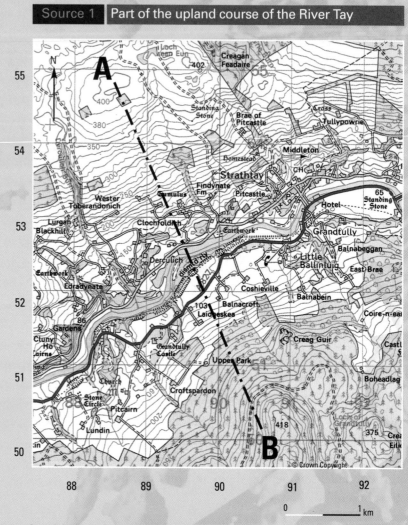

Source 1 Part of the upland course of the River Tay

In the lowlands

Source 2b is the valley cross-section near the sea. The flat and low-lying land is the floodplain. It is 0.6 km wide where the tributary River Earn meets the main River Tay.

Source 2 Cross-sections across the Tay valley

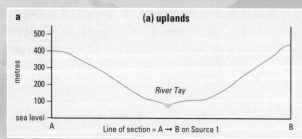

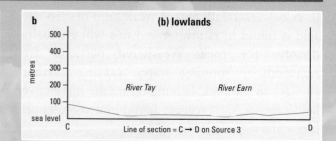

Source 3 Part of the floodplain and estuary of the River Tay

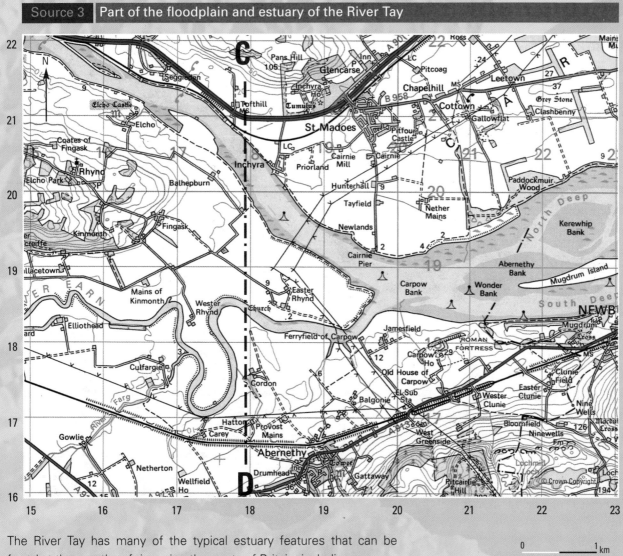

The River Tay has many of the typical estuary features that can be found at the mouths of rivers in other parts of Britain, including:

- a wide channel – up to 2 km
- sand and mud banks – Abernethy Bank
- some areas of marsh – in square 2119
- channels of deeper water – North Deep.

On reaching this tidal part of the river, a river's flow is reduced. The river loses much of its energy to transport its load and deposits it as sand and mud. These estuaries are difficult to navigate and to bridge, but they are good wildlife habitats. The River Earn has many features of a river in the lowlands (Source 4). Notice the big meander loop through squares 1718 and 1717 (Source 3). The black dashes marked around its edges show the levees. Settlements are larger than those in the uplands, but the danger of flooding means that they were carefully sited towards the edge of the floodplain where the risk of flooding was lower. The location of Abernethy is an example of this.

Source 4 Floodplain of the River Earn

Floodplain and delta
the River Ganges, India

The source of the River Ganges is in the Himalayas. The Ganges flows across its floodplain for over 1500 km through northern India. The Ganges delta is found at the mouth of the river as it flows into the Bay of Bengal. The Indian city of Calcutta lies on the western side of the delta, but most of the delta is in Bangladesh (Source 1).

Within the zone covered by the floodplain and delta of the Ganges live some 10 per cent of the total world population. This area is one of the most densely populated parts of the world. Most people living here are farmers for whom rice is the main food crop.

The **monsoon** climate brings summer rain, which fills up the River Ganges (see Source 1, page 8). For centuries the Ganges flooded the land around it between July and October. Each flood left another layer of fertile silt so that there is a great thickness of very rich and easy to work alluvial soils. Today a number of dams control the flooding of the river and supply water for winter crops. There is a neat landscape of tiny fields from which the farmers try to gain the highest possible output by hard work and by using high yielding varieties of rice, wheat and maize seeds (Source 2).

Source 1 | The River Ganges

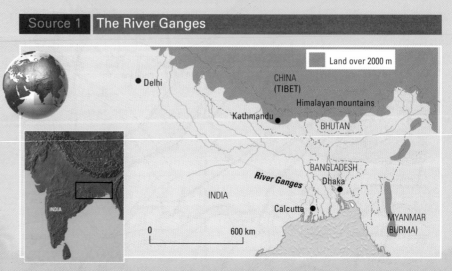

Source 2 | The Ganges floodplain is intensively farmed

Fact File | The Ganges delta

The delta is an example of a landform of river deposition. The physical features of the delta are labelled on the map and the reasons for the formation of the delta are written below.

Formation of a delta

1 The Ganges carries a large load of sediment.

2 The flow is slowed down by meeting the denser sea water.

3 Sediment is deposited faster than the tides can remove it.

4 River flow is blocked by so much deposition that the river splits up into distributaries.

5 Distributaries deposit sediment over a wide area, extending new land into the sea.

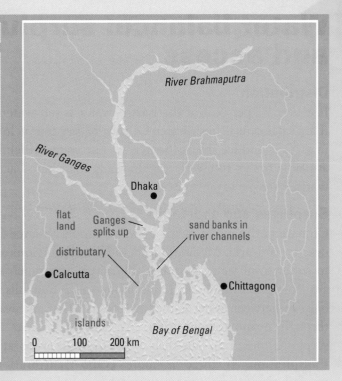

The *advantage* of the Ganges delta to the people of Bangladesh is that the land is very fertile. The silty soils are also easy to work. All the land is low lying (Source 3). There is water in the channels for irrigation during the dry season so that two or three crops can be grown. Many people can be fed each year.

The *disadvantage* is that the Bangladeshis live in a hazardous physical environment. The risk of being flooded is always present and many people have died. There are so many wide rivers to cross that land transport is slow and difficult. Access for ships is not easy because many channels are blocked by sand and mud. A wet environment in the tropics is a breeding ground for many diseases. Perhaps worst of all, Bangladesh is at the mercy of people living further up river. Nepal has cleared a lot of its forests; the higher rates of runoff in the mountains near the source have increased the severity of floods in Bangladesh at the river's mouth. India also uses the river for disposal of its waste products.

Source 3 | View of the Ganges delta

1.8 Water balance: surplus, shortage and access

Distribution

Just over 97 per cent of the Earth's water is salt water, so less than 3 per cent is fresh water. Of this, 77 per cent is locked up in glaciers and ice sheets (Source 1). The water which is easily available for us to use is very unevenly distributed across the world.

Surpluses and shortages

Climate is the major factor in water supply (Source 2), with desert, semi-arid, Mediterranean and Tundra regions all likely to suffer from water shortages. However, how much water is available is not simply the result of how much rain falls, but the balance between water gained by precipitation and water lost through **evapo-transpiration**. An area will have a water surplus if more rain falls than is lost through evaporation and transpiration. If there is more evapo-transpiration than precipitation, a water shortage can happen. In extreme cases, this imbalance causes flooding or drought.

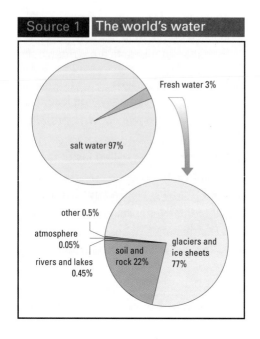

Source 1 | The world's water

Fresh water 3%

salt water 97%

other 0.5%

atmosphere 0.05%

rivers and lakes 0.45%

soil and rock 22%

glaciers and ice sheets 77%

Water is not always found where it is most useful to people. Areas of surplus are often located in remote, mountainous regions with high annual rainfall totals. Water collected and stored here has to be piped to where it is needed. In more arid areas, rain may fall just a few times a year or in one short rainy season, often in very heavy bursts. It can be very difficult to capture and store, and much of it is lost via surface runoff.

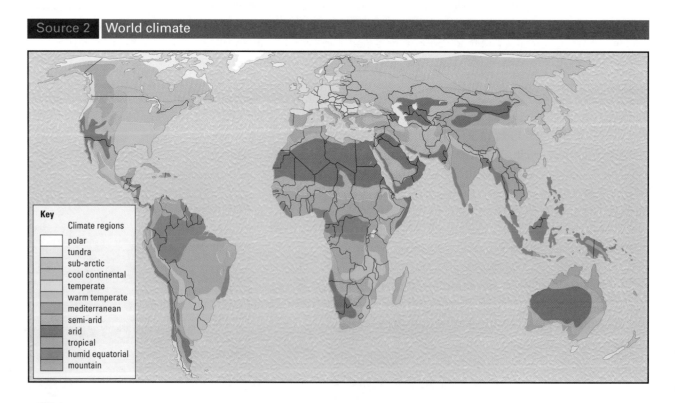

Source 2 | World climate

Key

Climate regions
- polar
- tundra
- sub-arctic
- cool continental
- temperate
- warm temperate
- mediterranean
- semi-arid
- arid
- tropical
- humid equatorial
- mountain

Source 3	Access to clean water

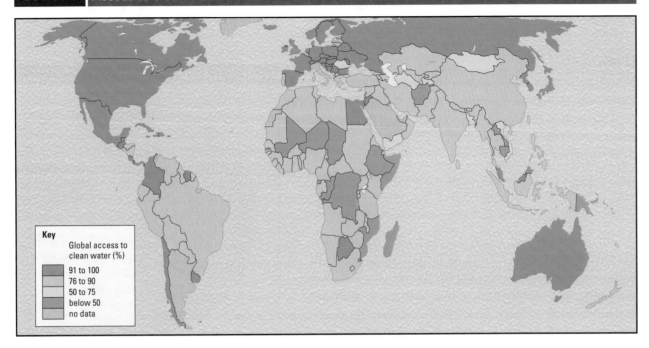

Key

Global access to clean water (%)

- 91 to 100
- 76 to 90
- 50 to 75
- below 50
- no data

Access to water

Despite the fact that 70 per cent of the Earth's surface is water, over 1.1 billion people today (a sixth of the population) do not have access to clean, safe drinking water (Source 3). It is estimated that by 2025 over 65 per cent of the world's population could be suffering from water shortages. Globally there is enough water for everyone, but there are huge differences in the amount of water people in different countries use. Source 4 shows average annual water use for a number of selected countries.

Differences in access to clean water are not just the result of climatic factors or imbalances within the hydrological cycle. Location, i.e. living in a rural or urban area, and wealth, i.e. living in an MEDC or LEDC, are also major influences. In rural areas in LEDCs often the only source of water is a local river or lake. Such sources are often contaminated by human and animal waste and chemicals from industry and farming. In urban areas, although water supplies are usually piped, the rapid growth of towns and cities is putting increasing pressure on water supplies. Access to proper sanitation is also a major problem. Half the diseases globally are related to

contaminated water or poor sanitation – mainly in LEDCs where people drink untreated, dirty water and untreated sewage goes into rivers or on the streets.

Where a country relies on a major river for water, problems can be caused if it flows through another country. The river may be dammed and diverted or water extracted and contaminated upstream, creating shortages or difficulties for those living downstream – shortages they can do nothing about. There is likely to be an increasing number of disputes about access to such supplies in future, even leading to 'water wars'.

Source 4	Average annual water use per person in selected countries (in thousand litres)

Australia	496
USA	215
France	102
Brazil	67
UK	40
Netherlands	25
China	20
Kenya	13
The Gambia	1

Water uses, management and supply

Uses

Water is essential to life, so it is one of the planet's most important resources. At home we use it for drinking, cooking and washing, but by far the biggest uses for water are farming and industry (Source 1). Almost three-quarters of global water use is for farming – growing and irrigating crops and providing for livestock. Industry uses twice as much as we do in the home – from generating power to manufacturing goods.

As the population of the world increases, so does the demand for water for a range of purposes. Advances in technology mean that people living in MEDCs especially use more and more water in washing machines, dishwashers and other modern domestic appliances. In LEDCs, as quality of life improves, more people have access to mains water, fuelling the need for more water.

| Source 1 | World water use |

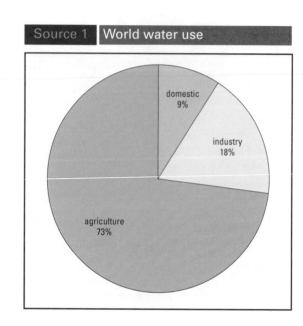

| Source 2 | The Colorado is one of the most controlled rivers in the world |

Increasingly water is used for recreational purposes for sports such as sailing, swimming and fishing. Large quantities are used in gardens, swimming pools and golf courses, especially in arid regions. The use of water in such regions cannot be sustainable – it is being used up at a far faster rate than it can be supplied. The desert cities of Las Vegas, Phoenix and Tucson in the USA, for example, rely heavily on water from the Colorado, the only river in this part of the country. In places the Colorado barely flows at all, with dozens of dams along its course and high levels of extraction (Source 2).

Management

Although the UK receives quite high levels of rainfall each year, there is limited capacity to store it when it does fall. The areas of heaviest rainfall are the highlands in the north and west – but the majority of the population live in the flatter and drier south and east. A number of major **reservoirs** have been built in highland areas where valleys can be dammed to create reservoirs. Water is then piped long distances to where it is needed.

In England and Wales the supply of clean water and the disposal and treatment of waste water has been the responsibility of ten major water companies since the industry was privatised in 1989 (Source 3). Water extracted from rivers, **reservoirs**, **aquifers** and **boreholes** and **waste water** collected via mains sewers needs to go through a number of different processes. These include screening, filtering and cleaning. Lastly, before it is piped through water mains to homes and industry, it is tested for quality.

DEFRA and the Environment Agency are the two bodies in the UK responsible for water supply. DEFRA is the government department responsible for policy-making, whilst the Environment Agency oversees the use and management of water resources, monitoring water quality and issuing abstraction licenses.

| Source 3 | The water companies and regions of England and Wales |

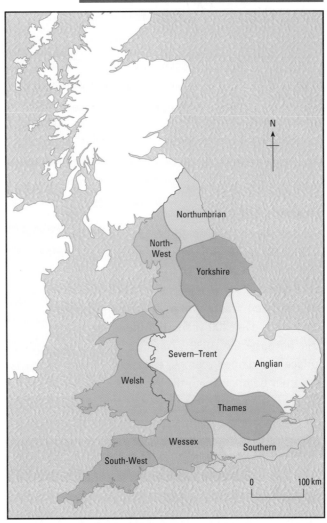

| Source 4 | A water treatment works |

Water supply
Thames Water

Thames Water is the largest water company in the UK and the third largest in the world. Nearly 7,000 people are employed serving 70 million customers worldwide, supplying water and undertaking projects not just in the UK, but in countries including Indonesia, Thailand, China, Australia and India.

Thames UK

Thames Water supplies 25 per cent of the population of England and Wales (see Source 3, page 23). It has 13 million customers in London and the Thames Valley and a further 1.5 million in Wales, providing clean water and removing **waste water**. It has supplied water to London for over 400 years, but the present company was privatised in 1989. Water is taken from both surface and ground sources, especially from rivers but also from boreholes into the chalk rock aquifer of the London Basin. It operates a number of treatment works in the region.

One of its first successes was to clean up the River Thames itself. For centuries the river was badly polluted and contaminated by industrial and human waste – it was biologically dead in places. Today it is one of the cleanest urban rivers in the world. Salmon is just one of the 120 species of fish found in the river.

During the 1990s Thames Water invested £250 million in constructing the London Ring Main. It is 80 kilometers long, 2.5 metres in diameter and between 30 and 60 metres below ground. Its twelve pumping stations are computer-controlled from Hampton, West London, providing London with a reliable water supply and moving water around where and when it is most needed.

| Source 1 | The River Thames is now one of the cleanest 'city' rivers in the world |

| Source 2 | One of the few places where Londoners can see evidence of the Ring Main |

Source 3 | Collecting water from a standpipe in Jakarta

The Marunda Project, Jakarta

In 1997 Thames Water began a project in Marunda in Indonesia's capital city, Jakarta, under the concession name of TPJ (Thames Pam Jaya). Jakarta is one of the world's fastest growing cities, having grown from half a million in 1930 to 11 million today. This rapid growth has placed great pressure on housing and services like water and sanitation. Many people live in shanty town areas called **kampongs**. In 1984 the 12,000 people living in one of these – Marunda – were moved to a new area in the north east of the city when their land was needed to develop the port.

The promise of a piped water supply failed to materialise and the only sources of water were through stand pipes, which meant queuing from 3.00 a.m. every day, or buying expensive bottled or canned water from local water vendors and tankers. Now, Thames Water supplies the 1,600 homes in Marunda with clean water direct from the mains as part of a 25-year contract. The cost is around 10p for a thousand litres, about a third of what people were previously paying vendors for limited supplies. This has improved the quality of life of the people in Marunda, reduced the risk of disease and enabled them to spend money on other basic needs. Many local people were employed to help during the construction stage of the project.

Source 4 | Marunda's people now enjoy a mains water supply

River management in a LEDC
the River Ganges, India

Three problems make effective management of the River Ganges difficult to achieve.

- **Climatic:** About 80 per cent of the yearly rainfall which supplies the Ganges falls in just four months of the year. As Source 1 shows, the Ganges suffers either from low flow (too little water for everyone's needs) or high flow (too much water causing flooding).
- **Political:** The River Ganges flows through three countries, which do not share the same interests. The source of the River Ganges is in Nepal; what happens there has knock-on effects for both India and Bangladesh. For example, removing forests from valley sides in Nepal increases surface runoff, which also increases flood risks in India and Bangladesh downstream.
- **Economic and social:** All three countries are relatively poor. They do not have unlimited amounts of money to spend on river management. There have been decades of high population growth and increased demands for food from the land.

| Source 1 | Management issues and methods along the Ganges |

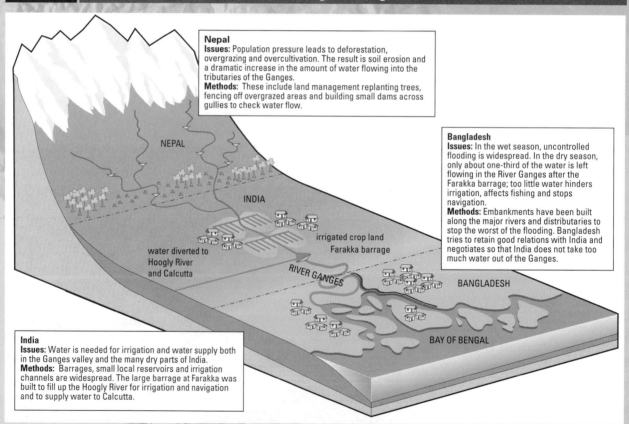

Nepal
Issues: Population pressure leads to deforestation, overgrazing and overcultivation. The result is soil erosion and a dramatic increase in the amount of water flowing into the tributaries of the Ganges.
Methods: These include land management replanting trees, fencing off overgrazed areas and building small dams across gullies to check water flow.

NEPAL

INDIA

Bangladesh
Issues: In the wet season, uncontrolled flooding is widespread. In the dry season, only about one-third of the water is left flowing in the River Ganges after the Farakka barrage; too little water hinders irrigation, affects fishing and stops navigation.
Methods: Embankments have been built along the major rivers and distributaries to stop the worst of the flooding. Bangladesh tries to retain good relations with India and negotiates so that India does not take too much water out of the Ganges.

water diverted to Hoogly River and Calcutta

irrigated crop land
Farakka barrage

RIVER GANGES

BANGLADESH

BAY OF BENGAL

India
Issues: Water is needed for irrigation and water supply both in the Ganges valley and the many dry parts of India.
Methods: Barrages, small local reservoirs and irrigation channels are widespread. The large barrage at Farakka was built to fill up the Hoogly River for irrigation and navigation and to supply water to Calcutta.

The floods are vital to the prosperity of most farmers in Bangladesh – they add another layer of fertile silt and water the land. However, how can floods be stopped from increasing beyond normal levels and ruining crops? This is almost impossible. Many of the flood protection schemes in Bangladesh are in urgent need of repair. Many embankments are of little use. Some have been washed away and others have merely diverted the water to other parts of the country.

River management in a MEDC 1.12
the River Mississippi, USA

The Mississippi is a major river system, draining most of the USA (Source 1). The Central Plains are an attractive area for agriculture and settlement, but the flood risk from the Mississippi has always been high as its flood record shows (Source 2). After every big flood, protection measures have been increased in size and numbers (Source 4).

- Everywhere levees, the high embankments on the river sides, have been increased in height, so that in places they are 15 metres high.
- Artificial channels have been cut across the necks of meanders so that the water flows straighter and quicker.
- Spillways have been built to take away excess water during floods.
- Over 100 dams have been built to control water flow on tributary rivers such as the Missouri, Ohio and Tennessee.
- Slopes have been afforested to reduce surface runoff.

Source 2	The flood record of the Mississippi

1900	Flooding from Cairo to the Gulf of Mexico
1927	Flooding up to 150 km wide around the river
1937	Flooding of an area the size of Scotland
1993	Flooding of almost everywhere north of Cairo

Source 3	Causes and consequences of the 1993 flood

THE MISSISSIPPI FLOODS OF 1993

It rained all spring and summer in the upper Mississippi and Missouri river valleys. 'Boy did it rain'. In some places it rained every day for two months. Most places had two to three times the usual amount of rain. There was nowhere for the water to go except into the rivers. The ground was waterlogged. By July an area the size of England was under water. Over 50,000 people were forced out of their homes. Many people were put up in schools and public buildings, which were used as temporary shelters. Nearly 50 people died. Crop losses were estimated at over $6 million and damage to property was put at $10 million.

Source 1	The drainage basin of the Mississippi

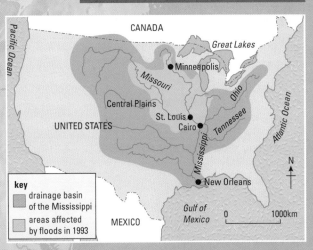

key
- drainage basin of the Mississippi
- areas affected by floods in 1993

Only a big and rich country could afford this amount of flood protection. The scale is enormous; for example, the levees stretch more than 3000 kilometres.

Considering the amount spent on flood protection measures, the size and scale of the floods in 1993 came as a big shock. People living in cities on the river banks and farmers in the Corn Belt states believed that they were no longer at risk from river floods. In 1993 it was the upper rather than the lower basin that was attacked, where flood protection was less. However, what these floods showed is that it is impossible to give a guarantee that a big river can be stopped from flooding, even in a wealthy MEDC.

Source 4	Methods of flood protection along the Mississippi

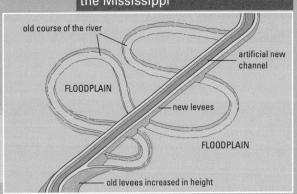

Hydro-electric power
the Three Gorges Dam, China

Most large dams are multi-purpose; there is more than one reason for building them. As they are so expensive to build, the cost can only be justified if several advantages result from their construction. The following are some of the advantages of building large dams:

- electricity production and power supply
- water supply for irrigation, homes and factories
- flood control
- improving navigation.

The Three Gorges Dam is being built mainly for electricity production and flood control.

Source 1 shows the location of the Three Gorges Dam, in the centre of China along the country's largest river, the Yangtze. Damming of the river began in 1997. When finished in about 2010, the dam will be the largest in the world. It will be 185 metres high and almost 2 kilometres wide. The lake, which will become the reservoir behind the dam, is expected to be up to 600 kilometres long. The plan is to build 26 generators producing over 18 000 megawatts of electricity, about 10 per cent of China's present energy needs. The electricity will be distributed over a wide area in central China. It is more than enough for the energy needs of ten big cities.

The major disadvantage of building such a massive reservoir is that the homes of more than 1 million people will have to be flooded by the rising waters behind the dam. These people need to be resettled. When a valley floor is flooded, the best and most fertile farmland is lost. The original plan was to rebuild villages and towns further up the valley sides. Here the slopes are steeper and without a covering of fertile silt.

There are great differences of opinion about whether the Three Gorges Dam Scheme should have gone ahead. The government of China sees it as an important symbol of China's modernisation. It is helping to reduce global pollution by cutting down on the burning of dirty coal in favour of clean and environmentally-friendly HEP. There is international pressure for China to reduce its emissions of greenhouse gases. It is the world's second largest burner of coal after the USA. At the same time, millions of Chinese will be protected from the risk of future floods from the Yangtze river.

Source 1	The Three Gorges Dam

Source 2 | The Chinese government viewpoint

Source 3 | Landscape in the Yangtze valley near to where the Three Gorges Dam is being built

- The scheme is vital for national needs.
- China cannot hope to industrialise further without HEP power from this dam.
- Coal burning in China will be reduced by 40 million tonnes a year.
- Emissions of carbon dioxide into the atmosphere will fall by 120 million tonnes each year.

Environmentalists argue that a series of smaller hydro-electric dams on the Yangtze tributaries would have been a more efficient way of generating power and managing the flood-prone river. They say that silt will be trapped behind the dam, making farmland lower down the Yangtze less fertile over time. Tonnes of industrial and human waste could be trapped behind the dam. Millions will die if the dam collapses. It is located in an area that has suffered earthquakes reaching 6 on the Richter Scale. Resettling people on steeper and poorer land will increase the dangers of soil erosion.

Source 4 | An artist's impression of the Three Gorges Dam when finished

China is keen to develop economically. More energy is needed. Electricity from water power is much cleaner than that made from fossil fuels such as coal and oil. However, smaller HEP schemes cause less environmental damage. The Chinese government has been forced to admit that the resettlement schemes have not gone as well as they had hoped. Many people already resettled have been unable to find employment in the new towns to which they have been moved. People living next to a dam are not the ones who usually benefit from the electricity it produces.

Source 1 | Immediate effects of river floods

River flooding

Flooding greatly increases a river's energy so that it can do more work. In times of flood, the deeper and faster river increases its load dramatically. Most rivers turn brown when in flood because of the large amount of material they are carrying in suspension. The amount of erosion by hydraulic action and abrasion is greatly increased. Many of the valley landforms in lowland areas have been formed by flooding. Levees and floodplains are formed by deposition after the river overflows its banks. Ox-bow lakes are cut off when the force of the river enables it to break through the meander neck.

River floods can cause a lot of damage. As with all the natural hazards, there are both immediate and long-term effects. The immediate effects of a river flood include loss of life, destruction of property and crops, and the disruption of communications (Source 1). The consequences can be serious. Many lives have been lost in floods along the big rivers in China – the Huang Ho is called 'China's sorrow' because its devastating floods have killed thousands of people. During the Rhine floods in early 1995, people in some Dutch villages had to be evacuated and the river current was too strong for barges to use the river safely.

In the longer term there is the cost of replacing what has been lost and damaged. In rich countries the risks may be covered by insurance. The poor in the less economically developed countries, however, may lose everything. With crop land ruined and animals lost, widespread famine can result. The need for emergency food aid in these circumstances becomes urgent.

Responses to floods

Flood protection measures are in operation along most big rivers. Where the costs of damage from flooding could be enormous, the scheme may be big and expensive such as the Thames Barrier which protects the city of London (Source 2).

Along most rivers the aim is to alter the channel or banks so that the river can carry more water without flooding. The most common measure of flood protection is to increase the height of the natural levees along the banks. This makes the channel deeper so that it can hold more water. Another measure that is often taken is to clean out the channel by dredging to make the river flow more efficiently.

On a much larger scale is the building of dams. In times of great runoff dams hold back water, and the river discharge in the area below the dam can be controlled. Dams can be a very effective measure of flood control but they are an expensive solution. They are often built as part of a scheme of **river basin management**. Not only is there flood control, but the water stored behind the dam can be used to supply houses and factories, irrigate crops and for recreation and shipping, as well as for HEP (hydro-

electric power). This is called a **multi-purpose scheme**. The River Rhône in France is an example of a managed river (Source 3).

Often, however, prevention is better than cure. By cutting down trees and by ploughing up and down slopes people have increased the rate of runoff in some areas, such as on the sides of the Himalayas in Nepal. The frequency and scale of flooding have been increased as a result. More care and attention given to the way in which the land in a drainage basin is used would reduce the risks of floods occurring in the first place.

Source 3 | River Rhône: river basin management

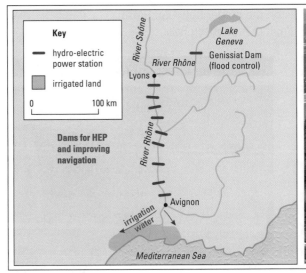

Key
— hydro-electric power station
▨ irrigated land
0 —— 100 km

River Saône
River Rhône
Lake Geneva
Genissiat Dam (flood control)
Lyons
Dams for HEP and improving navigation
River Rhône
Avignon
irrigation water
Mediterranean Sea

hydro-electric power station

river banks strengthened against flooding

canal for boats to use

1 a Draw a simplified labelled diagram of the hydrological cycle.
  b Explain the differences between each of the following:
    i evaporation and precipitation
    ii surface runoff and groundwater flow
    iii interception and percolation.
  c Explain why trees and rocks decrease surface runoff.

2 The values below give the average amount of water in a European river at different times of the year.

Months	J	F	M	A	M	J	J	A	S	O	N	D
Amount of water (cubic m)	500	450	620	700	750	820	900	840	650	580	550	530

  a Draw a bar graph to show the monthly amounts of water in the river.
  b State the difference in amount of water between the months with the largest and smallest amounts. Show your working.
  c Name the season of the year in which this river is most likely to flood.
  d The source for this river is high in the mountains. How does this help to explain the season in which the river is likely to flood?

3 a Describe four main ways rivers transport material.
  b Explain the differences between:
    i abrasion and attrition
    ii hydraulic action and solution.
  c How does a river change when it leaves an upland area and reaches a lowland area?
  d Why is deposition greater in lowland sections of rivers?
  e For a named example, explain how a major river delta formed.

4 Source 1 is a map of the Iguaçu Falls on the border between Brazil and Argentina. Source 2 is a photograph of the Iguaçu Falls.

Source 1 | Map of the Iguaçu Falls

Source 2 | The Iguaçu Falls

Key
Difference in height between X and Y is 82 metres
area covered by photograph
— · — international border
road
airport
hotel
Iguaçu National Park, southern Brazil

0 2 4 6 8 10 km

a Look at Source 1.
  i Name two facilities provided for tourist visitors to the area.
  ii State the height of the Iguaçu Falls.
b In which direction was the camera pointing when the photograph in Source 2 was taken?
c Describe the attractions of the area for tourist visitors.
d Suggest reasons why the site marked Z in Source 1 could make a good site for a hydro-electric power station.
e Look at Source 2. Make a frame and draw a labelled sketch to show the physical features of the area shown in the photograph.
f Explain how the features shown in Source 2 are formed by river erosion.

5 a What are the main differences between mechanical and chemical weathering?
  b Explain the processes at work in:
    i freeze-thaw
    ii exfoliation
    iii the formation of limestone pavement.

6 a Use the OS map on page 16 to answer the following:
    i Describe the changes in relief (shape of the land) you would see if you walked from point A to point B.
    ii What is the area of land used for coniferous woodland (in square km)?
    iii Name five different types of land use in square 9153.
    iv Name the river feature at 910528.
  b Use the OS map on page 17 to answer the following:
    i Draw a sketch map of the River Tay estuary east of easting 20. Show and label four physical features commonly found in a river estuary.
    ii What are the advantages and disadvantages of this estuary for shipping?

7 a What is meant by:
    i water balance?
    ii water deficit?
    iii water surplus?
  b List the main ways in which water is used. Try to group your answers under suitable headings.
  c Global water use is increasingly rapidly. Explain why.
  d For both the Thames Water Ring Main in London, UK and the Marunda Project in Jakarta, Indonesia, describe the scheme and the benefits it has brought.

8 a Use either page 26, the Ganges, or page 27, the Mississippi. Write your own case study notes in the following sections:
    i Location of the river (and tributaries)
    ii Flood risks
    iii The main ways in which the river is managed.
  b The Three Gorges Dam in China (pages 28–9) is a multi-purpose scheme.
    i What does this mean?
    ii Describe the different parts of the scheme.
    iii Summarise the arguments for and against the scheme.

Hazards

Unit Contents

A survivor of the 1995 Kobe earthquake. What are the short and long term effects of such natural disasters?

Natural hazards

There are two main types of natural hazards – those caused by tectonic activity and those caused by extreme weather events. Tectonic hazards include earthquakes (pages 38–9), tsunamis and volcanoes (pages 42–3). Weather hazards include flooding (page 53), drought and hurricanes (pages 48–9). Such hazards are usually found in specific regions across the world, but exactly when and where they may occur is not easy to predict.

Source 1	The world's tectonic plates

Eurasian plate

North American plate

Pacific plate

Pacific plate

Philippine plate

Nazca plate

African plate

Indo-Australian plate

South American plate

Antarctic plate

Antarctic plate

Key

constructive (divergent) plate margin

▲▲▲▲ destructive (convergent) plate margin

→ direction of plate movement

Plate tectonics

The crust of the earth is made up of a number of **tectonic plates** (Source 1). These plates move over the surface of the globe. When two plates are moving apart, for example in the oceans, the margin between them is called a **constructive plate margin**. It is called this because new crust is being created, for example, along the mid-ocean ridge in the Atlantic Ocean. When two plates move towards each other, like the Nazca plate and the South American plate, the margin between them is called a **destructive plate margin**.

Tectonic plates shape the landscape by creating new rocks and forming mountains and valleys by processes such as folding and faulting (Source 2). They also cause a range of natural hazards, including earthquakes, tsunamis and volcanoes –

Source 2	Tectonic forces push up mountains

The volcanic peak of Mount Rainier rises above the folded Cascade Mountains in North-West USA where the North American and Pacific plates meet.

the most powerful natural forces on the planet. The majority of mountain building and tectonic hazards occur along or close to plate boundaries (Source 1).

Forces beneath the crust

It is hot beneath the surface of the earth. The heat is so great that the rocks below the crust are molten. This molten rock is called **magma**.

At particular places, called 'hot spots', magma rises to the surface in a series of **convection currents**. When the currents reach just below the crust they can rise no more. Instead, the hot currents spread out and carry the crust in a series of plates across the globe, as shown in Source 4.

As the convection currents spread out they pull the crust apart (A in Source 4). At this point a crack in the crust is formed, called a **rift valley**. Molten magma reaches the surface through this crack, producing volcanoes and, under the ocean, **ocean ridges**. Here, new crust is being constructed, hence the name 'constructive plate margin'.

Convection currents also move the crust towards other plates. One of the plates will be forced down beneath the other at the **subduction zone** (B in Source 4). The rocks left on the other plate are crumpled into high **fold mountains**. The crust which has been dragged beneath the surface is destroyed – hence these margins are called 'destructive plate margins'. In fact some of the lost crust is heated up to form new magma which eventually finds its way back to the earth's surface.

| Source 3 | Satellite view of part of the East African Rift Valley – tectonic forces are tearing apart the earth's crust |

Wherever the tectonic plates are being pulled apart or crushed together, earthquakes can occur. Shock waves travel through the rocks causing violent movements of the land which destroy settlements and may also lead to loss of life.

Sometimes the plates pass alongside each other, neither colliding nor pulling apart. These plate margins are known as **conservative plate margins**, since crustal material is neither being created nor lost. However, the crust may crack and faults like San Andreas will be created, along which other earthquakes occur like that in San Francisco in 1906.

| Source 4 | Forces beneath the earth's crust produce mountains, volcanoes and ocean ridges |

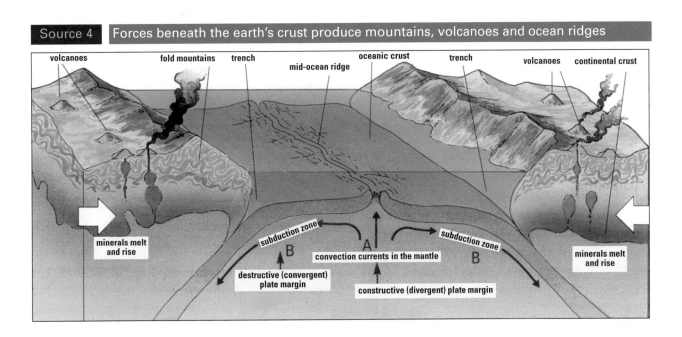

Faulting and earthquakes

Faults

The earth's tectonic plates are constantly moving, causing the crust to crack, forming faults. When the land along a fault moves, earth tremors or earthquakes may occur. This is especially common when land on one side of a fault is sliding past the land on the other side, as in the San Andreas fault (Source 1). Typically the rocks either side of the fault move very slowly, just millimetres at a time. Sometimes the rocks stick together, building up pressure until eventually the rocks tear apart, causing an earthquake.

Thousands of earthquakes occur every year, but most can only be detected by instruments called seismographs. Up to 50 major earthquakes occur each year, and these can cause widespread damage and destruction, especially if they happen in densely populated areas.

Source 2 shows what happens during an earthquake. The centre of the earthquake underground is called the **focus**. Shock waves travel outwards from the focus. These are strongest closest to the **epicentre** (the point on the surface directly above the focus). The amount of damage caused depends on the depth of the focus and the types of rocks. The worst damage occurs where the focus is closest to the surface and when rocks are soft.

| Source 1 | The San Andreas fault runs down the western side of California |

| Source 2 | Earthquake shock waves are measured by the Richter Scale which is used to determine the likely damage caused by an earthquake |

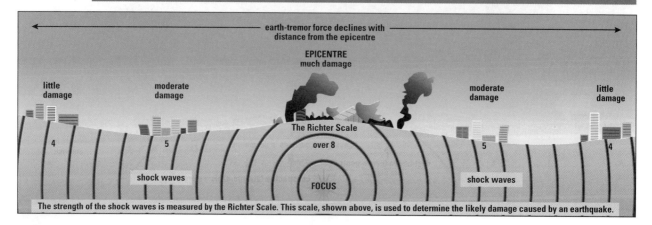

earth-tremor force declines with distance from the epicentre

EPICENTRE
much damage

little damage

moderate damage

The Richter Scale

moderate damage

little damage

4

5

over 8

5

4

shock waves

FOCUS

shock waves

The strength of the shock waves is measured by the Richter Scale. This scale, shown above, is used to determine the likely damage caused by an earthquake.

Tectonic hazards in a LEDC 2.3
Gujarat, India

At 8.46 a.m. on 26 January 2001, a 6.9 magnitude earthquake struck the state of Gujarat in north-west India (Source 1). The epicentre was 20 km north-east of the city of Bhuj, but its effects were felt in 21 of the state's 25 districts. Twenty thousand people were killed and over 150 000 injured, mainly by collapsing buildings. A major earthquake is expected every 30 years in this region which is located on the Indo-Australian tectonic plate where it collides with the Eurasian plate. However, it is almost impossible to predict exactly when and where a major quake will occur.

The earthquake lasted between 15 and 20 seconds and affected almost 16 million people in Gujarat – and was felt as far away as Bangladesh, Nepal and Pakistan. At its centre, the ancient walled city of Bhuj was almost totally destroyed, as was nearby Anjar. The high death toll was partly because of the poor construction of the buildings in the region, but also it was a public holiday and many people were at home.

In the short term, help came from Indian troops, from neighbouring Pakistan and from across the world. Many villages were difficult to reach, especially with transport, power and communication links badly damaged. This is a major problem when so many people are trapped under rubble with rescuers unable to reach them sometimes for several days. Hundreds of thousands of homes were damaged or destroyed. The government appealed for substantial financial aid from the World Bank (£650 million) and Asian Development Bank (£325 million) to help fund rebuilding.

In many areas the rubble from the earthquake still remains. However, in some villages a joint project between the Indian government and NGOs has seen the hundreds of new 'earthquake-proof' buildings constructed (Source 2). Six thousand of these are planned, built using local materials (including earthquake rubble for the foundations). Towns in the region are growing rapidly and not all new buildings are being built to this standard, mainly because of increased cost. Many people criticise the government for not forcing developers to make all new buildings earthquake-proof.

Source 1   The state of Gujarat in north-west India

INDIA

Bhuj ·

GUJARAT

Arabian Sea

0   200 km

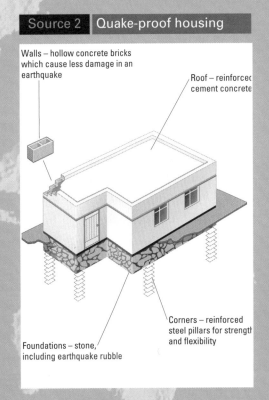

Source 2   Quake-proof housing

Walls – hollow concrete bricks which cause less damage in an earthquake

Roof – reinforced cement concrete

Corners – reinforced steel pillars for strength and flexibility

Foundations – stone, including earthquake rubble

Tectonic hazards in a MEDC
Kobe earthquake, 1995

Early in the morning on Tuesday, 17 January 1995 the shock waves of a huge earthquake roared through the city of Kobe. Measuring 7.2 on the Richter Scale, it was the worst earthquake to hit Japan for 50 years.

- 6432 people were killed.
- Over 100 000 buildings were destroyed.
- 300 000 people were made homeless.
- Rail links, bridges, the main expressways, docks and port area were badly damaged.
- The cost of the damage was estimated at $200 billion.
- Over 300 fires broke out destroying 7000 homes and responsible for 500 deaths devastating an area of 100 km^2 in central Kobe.

This earthquake was caused by the Philippine plate moving beneath the Eurasian plate (Source 1). The rocks had locked together many years ago and the pressure had built up each year since then. Suddenly they jerked free, and the shock waves were released. The destruction is shown in the photograph on page 35, and in Sources 1 and 2.

The epicentre of the earthquake was near Awaji Island. Here only buildings were destroyed. The greatest destruction was where most people live – in the cities of Kobe, Akashi and Ashiya. The famous bullet train tracks, motorways and bridges were all badly damaged. Broken gas pipes and electricity lines caused fires to rage throughout the built-up areas – especially among the many wooden houses built to withstand the shock waves.

Source 1 Cause and effect of the Kobe earthquake

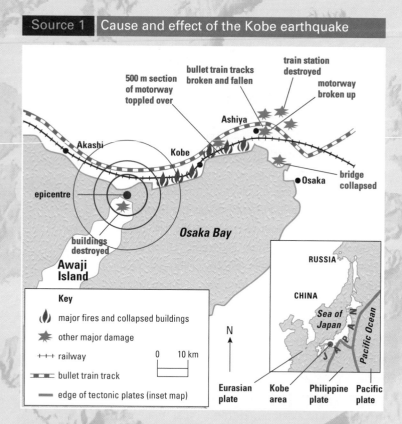

Source 2 The Kobe earthquake caused severe damage

Fires broke out consuming many destroyed buildings

Very few buildings left standing

Electricity and telephone lines brought down

The Kobe earthquake killed 6000 people, compared to 20 000 in Gujarat state, although Kobe was a 7.2 magnitude quake compared to 6.9 in Gujarat. One of the main reasons for the differences was that in Kobe many buildings and structures were built to withstand earthquakes, and the response time to help trapped and injured people was much quicker. LEDCs rarely have the resources to respond as well to such disasters in either the short or long term, or to plan and reduce the impact of earthquakes.

Japan experiences over 1000 earthquakes every year. Fortunately most are quite minor tremors, or occur deep underground or under the sea, and have little impact. However, the Japanese are very aware of the danger that major earthquakes bring and considerable time, effort and money is spent on trying to strengthen buildings and infrastructure, plan responses and predict future earthquakes. New buildings and structures have to be built to withstand strong tremors (Source 3). Regular earthquake drills are held in schools and places of work, and every year a full-scale practice for armed forces and emergency services is held. However, such action will only help reduce the impact of an earthquake, even in a rich country like Japan, especially if it occurs in a densely populated area.

Predicting earthquakes

Scientists know where earthquakes will strike – along the active plate margins. They find it much more difficult to say when they will happen. Before an earthquake the land may be seen to rise or tilt. Sometimes the water level in wells is seen to fall. If local people notice these changes they can alert everyone to reach places of safety, well away from buildings.

If these changes do not occur, or are not seen, there is very little chance of predicting earthquakes. There have been recent improvements in detecting changes in electrical signals and in registering radioactivity emissions. In order to register such changes many more scientific stations or satellites capable of recording these indicators are needed.

| Source 3 | Building to survive earthquakes |

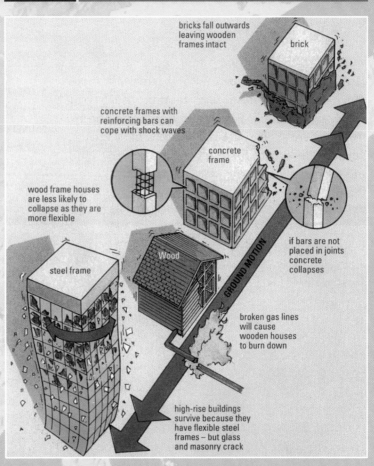

bricks fall outwards leaving wooden frames intact

brick

concrete frames with reinforcing bars can cope with shock waves

concrete frame

wood frame houses are less likely to collapse as they are more flexible

if bars are not placed in joints concrete collapses

Wood

steel frame

GROUND MOTION

broken gas lines will cause wooden houses to burn down

high-rise buildings survive because they have flexible steel frames – but glass and masonry crack

Resisting earthquakes

If scientists were able to predict when earthquakes are likely to happen, many lives would be saved. Even if earthquakes could be predicted accurately, they would still damage buildings. Recent earthquakes in different parts of the world have allowed town planners to form a picture of what types of buildings can resist earthquakes.

Source 3 shows how different building materials respond to shock waves.

- Wooden houses may burn in the aftermath of an earthquake, as they did in Kobe.
- Bricks fall out of buildings, so they are not good materials in earthquake zones.
- Concrete is much better, as long as it is reinforced by strong, flexible steel bars.
- High-rise buildings with flexible steel frames do survive, but falling glass and bricks can cause injury and death.

As Kobe was rebuilt after the earthquake, more safety measures were used in design and construction.

Volcanoes and volcanic activity

Volcanic activity is important right across the earth's surface. There are over 500 active volcanoes which still erupt from time to time. Each year there are around 30 or 40 eruptions. Some eruptions are slight, but others have serious effects, causing loss of life and damage to land and property.

Volcanic activity can be divided into two types:

- **intrusive volcanic activity** – where **lava** (molten rock) cools and solidifies beneath the surface
- **extrusive volcanic activity** – where lava reaches the earth's surface before it cools and solidifies.

Source 1 The world's active volcanoes

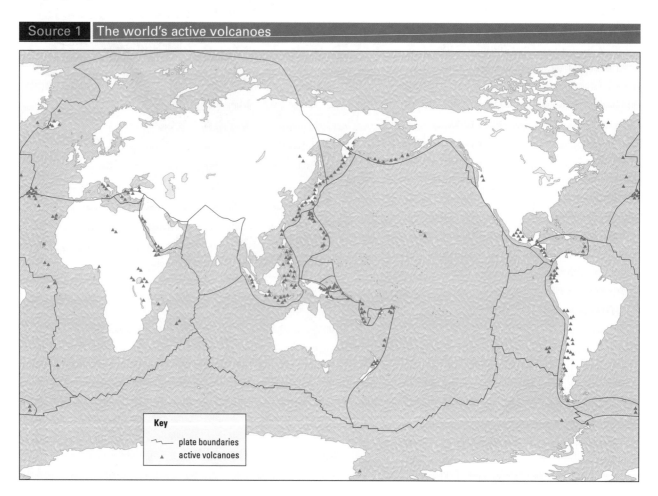

Key

plate boundaries

active volcanoes

Volcanoes are examples of extrusive volcanic activity, where magma from beneath the earth's crust reaches the surface as lava. Source 1 shows the location of the world's active volcanoes. The majority of these are found at destructive plate margins, where one tectonic plate is sliding down, or subducting, beneath another one (Source 4, page 37).

Many volcanoes are found around the edge of the Pacific Ocean – down the west coast of North and South America, through the Philippines, Indonesia and Japan. This circle of active volcanoes is called the Pacific Ring of Fire.

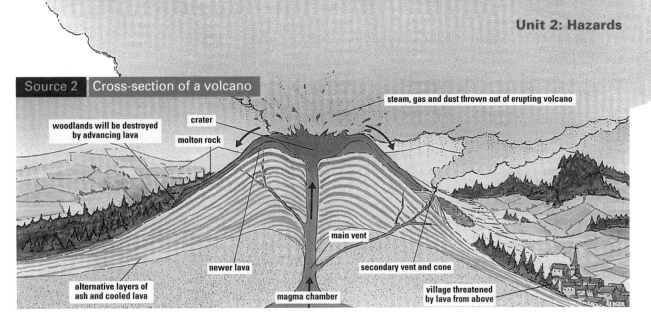

Source 2 | **Cross-section of a volcano**

steam, gas and dust thrown out of erupting volcano

woodlands will be destroyed by advancing lava

crater

molton rock

main vent

newer lava

secondary vent and cone

alternative layers of ash and cooled lava

magma chamber

village threatened by lava from above

Extrusive volcanic activity

A volcano is a crack in the earth's surface which provides an outlet for lava, steam and ash. The lava passes through a pipe called a vent from inside the crust to the earth's surface where it erupts (Source 2). After the eruption a crater is left on the surface. Lava which has erupted hardens on the surface.

The escape of lava from a volcano can take place under pressure, resulting in explosions of steam, gas and dust. These are hurled high in the air and molten rock pours down the volcano's sides.

Eruptions may damage property – houses, roads and farmland. The dust which reaches the upper layers of the atmosphere can cause climatic change on a global scale, blocking out the sun's heat and increasing rainfall. An example of a very destructive volcano is Mount Pinatubo in the Philippines (see Unit 2.6).

Volcanoes can also have positive effects. Ash and lava turn into fertile soils. Hot magma beneath the surface heats up underground water, providing steam for heating and power. Examples of these positive effects are described in Unit 2.7.

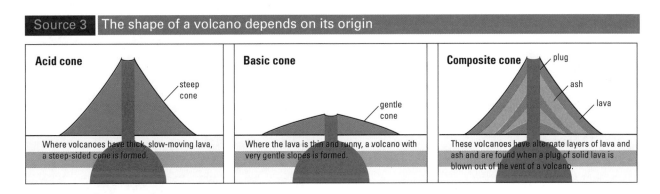

Source 3 | **The shape of a volcano depends on its origin**

Acid cone

steep cone

Where volcanoes have thick, slow-moving lava, a steep-sided cone is formed.

Basic cone

gentle cone

Where the lava is thin and runny, a volcano with very gentle slopes is formed.

Composite cone

plug

ash

lava

These volcanoes have alternate layers of lava and ash and are found when a plug of solid lava is blown out of the vent of a volcano.

There are many different kinds of lava. Some are very thick and 'sticky' (viscous). Others are thin and 'runny'. Lava which is made up of mainly acid minerals is viscous. When it flows out it solidifies quickly forming steep-sided **acid cones**.

Basic cones are formed by less acid lavas which flow across the landscape forming gentle slopes or, sometimes, flat plateaus.

Tectonic hazards in a LEDC
the eruption of Mount Pinatubo

Mount Pinatubo is a volcanic mountain located about 100 km north-west of Manila, the capital of the Philippines (Source 1). By June 1991, the volcano had been peaceful for more than six centuries. During this time the ash and lava from previous eruptions had weathered to become fertile soil which was used to cultivate rice. Then, suddenly, the volcano came to life (Source 2).

Advance warning that the volcano was about to erupt gave the authorities time to evacuate thousands of people from the nearby town of Angeles. Some 15 000 American airmen and women also left the nearby Clark air base. The level of activity increased and finally on 12 June the volcano sent a cloud of steam and ash some 30 km up into the atmosphere.

More deadly than the steam and ash were the **pyroclastic flows**. These are the burning gases that descend from the volcano at speeds of over 200 kilometres an hour. They engulf and burn everything in their path. It is the main killer of people and destroyer of wildlife in volcanic eruptions (see Unit 2 on Vesuvius).

Source 1 Mount Pinatubo in the Philippines

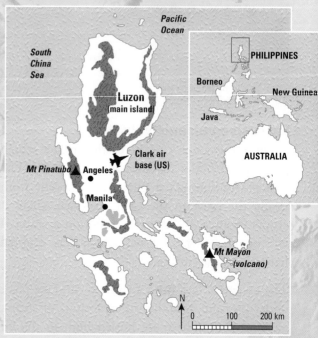

Source 2 Pinatubo erupts in June 1991 – a lorry tries to outrun the rapidly advancing pyroclastic flow

Fact File

- Ash fell to a depth of 50 cm near the volcano, and over a 600 km radius of the volcano it was still over 10 cm deep.
- The volume of ash in the atmosphere turned day to night and hampered the rescue operations.
- Torrential rain accompanied the eruption, and much of the ash was rained back to earth as mud, causing thousands of buildings to collapse under its weight.
- Power supplies were cut and roads and bridges were left unusable, as was the water supply which was quickly contaminated.
- Some 350 people were killed, mostly by pyroclastic flows.

When a volcano erupts much cinder, ash and lava falls on the slopes. The lava solidifies, and the cinder and ash remain in a loose and unstable condition on the upper slopes.

This is what happened after the 1991 eruption of Mount Pinatubo.

At a later date heavy rain washed this loose material down the volcano sides. This occurred after typhoons hit the Philippines in 1993 and 1995. The local people were inundated with mud avalanches – the **lahars** (Source 3).

| Source 3 | Filipinos flee the lahar, 5 September 1995 |

Boiling mud swamps towns

Philippines. Hundreds of Filipinos were plucked from the roofs of their houses yesterday after walls of boiling mud swamped towns.

Torrential rain from Tropical Storm Nina loosened a torrent of volcanic debris, called lahar, from Pinatubo volcano which swept into towns. More than 65 000 people were forced to flee the area.

A spokesperson for the regional disaster office said: 'We had already been hit by a 10 ft lahar last week. And then Bacolor got hit with another 10 feet of lahar today.'

Bacolor, a town of 20 000 people, was turned into a wasteland of mud and ash. Houses were buried, forcing residents to haul their beds, cooking pots and clothes to the roof.

Helicopters evacuated scores, but many refused to go, not wanting to leave their belongings behind.

The long-term effects

The effects of the eruption were felt long after the volcano became dormant again. With thousands of people living in refugee camps, malaria and diarrhoea quickly spread. Huge quantities of dust, some 20 million tonnes, were left in the atmosphere. Scientists believe that the dust has resulted in a lowering of average temperatures, and that it will delay global warming.

The cost to the Philippines was immense. Crops, roads and railways, business and personal property were destroyed, amounting to over $450 million.

Looking towards recovery

The Philippines is a LEDC and has little money to spend on rebuilding the part of Luzon devastated by the eruption.

The response of the authorities of Central Luzon needs to focus on:

- protecting against further lahars and flash flood damage by building dykes and dams
- establishing new work for the farmers and other workers, for example those working at Clark air base, well away from the danger area
- creating new towns and villages away from the danger area.

The great cost of these needs cannot be borne by the people of Luzon alone. Here is a good example of how the international aid agencies, such as OXFAM, can be used to provide development funds in LEDCs.

Living with hazards
volcanoes, Iceland and Southern Italy

Volcanoes are usually destructive. Eruptions often lead to loss of life and may also damage property and crops. However, volcanic activity can also have positive results.

Iceland

Iceland is situated in the North Atlantic Ocean. It is also located on a major plate boundary – the mid-Atlantic ridge (Source 1). As a result of this, there is a lot of volcanic activity on the island.

- It has over 200 volcanoes (Source 2).
- There are over 800 hot springs and geysers.
- Ten per cent of the land surface is lava fields.
- New land is being created as the two plates (the Eurasian and North American plate) separate along the ridge.

| Source 1 | Iceland on the mid-Atlantic ridge |

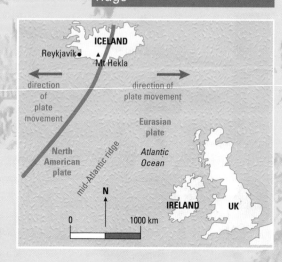

| Source 2 | Mount Hekla erupts in February 2000 |

| Source 3 | Bláa Lónið – the 'Blue Lagoon' near Reykjavik – Icelanders take advantage of these naturally heated waters which vary in temperature between 33°C and 45°C |

Volcanic activity in Iceland has some positive effects:

1 **Hot spring water** is carried by pipeline to Reykjavik, the capital of Iceland, giving most people a cheap and reliable form of energy which does not damage the environment. Much of the water used in the city today comes from a borehole 45 km away, which yields water at over 300°C. Hot water at a lower temperature provides recreational opportunities for Icelanders and tourists alike (Source 3).

2 **Electricity is generated** from the geothermal resources of hot water which lie under the ground in many parts of Iceland. Steam from the hot water deposits found underground is used to power turbines, which in turn generate electricity for factories and homes.

3 **Greenhouses** not far from Reykjavik are geothermally heated, allowing some Icelanders to produce vegetables, fruit and even flowers.

Source 4 Petrified body excavated from the ash at Pompeii

Source 5 Land use on the fertile soil surrounding Vesuvius

Southern Italy

In southern Italy, volcanic activity has brought different benefits. Over 1900 years ago, in AD 79, the volcano Vesuvius erupted. The town of Pompeii was completely destroyed. The eruption sent burning gases (pyroclastic flows) down the sides of the volcano killing the people of the town. Ash from this eruption, together with many smaller ones that have taken place since, has covered the landscape in this part of southern Italy. Source 4 shows the petrified remains of a child killed by pyroclastic flows from Pompeii in AD 79.

Over many years, the ash has weathered to produce fertile soils. This has led to many people settling around the volcano, where the soils are used to grow olives, fruit, vines and nuts as well as a variety of market garden produce (Sources 5 and 6). Villages and towns circle the higher, sterile slopes, where the lava has not yet weathered, so nothing can be grown. In other places where the ash and lava are loose, forests have been planted to stop slope erosion.

The slopes of Vesuvius, in spite of the ever-present danger of new volcanic eruptions, will always provide rich soils to encourage farmers to work the land.

Source 6 Mechanisation on the rich farmland surrounding Vesuvius

Predicting volcanic eruptions

Like Vesuvius, volcanoes may erupt only after hundreds or thousands of years of being quiet (**dormant**). So it is very difficult to predict eruptions. There are warning signs, however. Near the time of an eruption the magma beneath the volcano comes close to the surface. This will cause:

- the escape of gases, particularly sulphur dioxide, which can be monitored with special equipment

- a number of small earthquakes which can be measured with special equipment
- swelling of the sides of the volcano.

The problem is that monitoring equipment is very expensive and could be in place for generations without detecting any signs. Perhaps the best way is for people living near volcanoes to keep a regular watch on any changes which may indicate a coming eruption.

Weather hazards: tropical storms

Source 1 | Global distribution of tropical storms

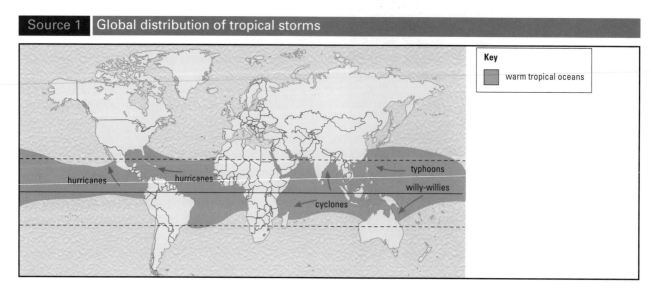

Tropical storms or cyclones are large areas of low air pressure which bring torrential rain and very strong winds to tropical regions. Source 1 shows their distribution and what they are called in different parts of the world. For example, severe tropical storms are called hurricanes in North and Central America and the North Atlantic. Depending on the location, there may be from six to over 20 each year, but most occur between mid-summer and early autumn when the water is warmest.

How do tropical storms form?

Tropical storms need warm water over 27º Celsius to form. The water heats the air above it, resulting in warm, moist air rising quickly upwards. As it rises, it turns and spins inwards (Source 2), creating an area of very low pressure in the centre or 'eye'. In the northern hemisphere, the storm winds rotate anti-clockwise; in the southern hemisphere they rotate clockwise. The rising air quickly cools down, forming thick, dense cumulo-nimbus clouds which bring very heavy rainfall.

Source 2 | Cross-section through a tropical storm

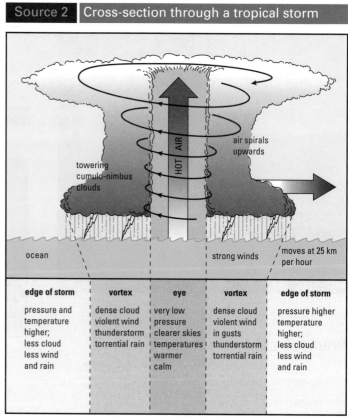

edge of storm	vortex	eye	vortex	edge of storm
pressure and temperature higher; less cloud less wind and rain	dense cloud violent wind thunderstorm torrential rain	very low pressure clearer skies temperatures warmer calm	dense cloud violent wind in gusts thunderstorm torrential rain	pressure higher temperature higher; less cloud less wind and rain

These masses of rotating low pressure can be over 100 km wide and travel at up to 50 kph. Inside wind speeds can reach over 250 kph around the edges of the central eye. The eye itself is calm. Source 3 is a satellite image showing the swirling mass of cloud and the clear 'eye' at the centre of a hurricane in the Atlantic Ocean off Florida and the USA. Tropical storms need warm water for energy – once they reach land they quickly lose power.

Classifying and naming tropical storms

Once wind speed in a tropical storm reaches 55 kph, it is classified as a hurricane, measured on the five-point Saffir-Simpson scale (Source 4).

Tropical storms are given names by meteorologists. These are from alphabetical lists, with alternate male and female names over a six-year cycle. There are different lists (and names) for different parts of the world. If a storm is particularly devastating, the name is usually retired and replaced. Names help to identify and track individual storms, especially as there may be more than one happening at a time.

| Source 3 | Satellite view of a tropical storm or hurricane |

| Source 4 | The Saffir-Simpson classification |

Category	Wind speed (kph)	Pressure (mb)	Storm surge (m)	Damage
1	119–153	<980	1.0–1.7	Minor – trees, mobile homes
2	154–177	979–965	1.8–2.6	Buildings – roof, windows Small boats from moorings Flooding
3	178–209	964–945	2.7–3.8	Structural Flooding over a metre up to 10 km inland
4	210–249	944–920	3.9–5.6	Major – destroys buildings, beaches, flooding up to 10 m and up to 5 km inland
5	<250	>920	Over 5.7	Catastrophic – destruction aup to 5 metres above sea level Mass evacuation needed

NB Tropical storms have wind speeds between 55 and 118 kph.

The effects of tropical storms

The average length of a tropical storm or hurricane is 10 days, but the biggest can last for up to four weeks. They cause three main types of damage – wind, storm surges (in coastal areas) and floods. Winds can destroy trees, crops, buildings, transport, power and communications. Storm surges along coastal areas can be devastating as huge waves hit the land. Torrential rainfall can last for several hours or several days, causing widespread flooding inland. This can trigger potentially deathly landslides and mudslides.

The long term impact of a tropical storm or hurricane often depends not just on its ferocity, but whether it hits an MEDC or LEDC. Whilst damage cannot be prevented, wealthier MEDCs can board-up properties and evacuate in time as warnings can be broadcast in advance of approaching storms monitored via satellite. They are also more able to cope with clearing up the damage and restoring business and economic activities after the event. LEDCs can be devastated by the effects of a tropical storm. With buildings and farmland ruined and little money available to rebuild, it can take years for the people and economy to recover. Even then, an LEDC may often be dependent on international aid both for long-term rebuilding and short- and mid-term help.

Hurricane Mitch

During October 1998, Central America was hit by its most destructive tropical storms and hurricanes for 200 years (Source 1). Mitch began as tropical depression on 21 October to the south of the Caribbean. A day later it became first a tropical storm, then a hurricane as wind speeds increased rapidly. By 26 October it had become a category 5 hurricane with speeds of over 250 kph, moving west across the Caribbean.

Mitch makes landfall

Whilst meteorologists could track Mitch via satellite, they could not accurately predict which direction it might eventually take. Even had they known where it would make landfall, very little could be done to protect the area, nor could people be evacuated easily in time. By 28 October, Mitch had started to move south-west towards Honduras (Source 2). Although wind speeds inside the hurricane were still high, they had started to fall. The main problem for Honduras and neighbouring Nicaragua and El Salvador was the relatively slow movement of the whole system. As a result of this, rainfall was intense and 180 cm fell in just three days.

The sheer volume of water created widescale flooding, destroying buildings, roads, bridges, crops and livestock. It also caused numerous mudslides which claimed a large number of victims. By the time Mitch turned north to Mexico, at least 10 000 people had lost their lives. Even then it had not finished as winds increased again before it reached Florida in the USA.

Source 1	Hurricane Mitch: record breaker

- Second longest-lasting category 5 hurricane (33 hours)
- Third longest continuous period of high winds (15 hours)
- Fourth strongest hurricane (winds of 249 kph)
- Fourth lowest air pressure ever measured (905 mbs)

Source 2	The trail of devastation made as Mitch tore through Central America

Key
← main track of Hurricane Mitch

Source 3	The effects of Mitch by country

Country	Dead/missing (approx)	Other
Honduras	14 000	2 million homeless
Nicaragua	3 000	$\frac{3}{4}$ million homeless
Costa Rica	7	3 000 evacuated
El Salvador	400	50 000 homeless
Guatemala	200	80 000 evacuated
Belize	0	10 000 evacuated
Mexico	6	Thousands evacuated

The impact on Central America

For most of the countries hit by Mitch, there was little warning nor anywhere to go for shelter. Two million people in Nicaragua were affected by Mitch (Source 3). Mudslides triggered by torrential rain destroyed villages, schools, health facilities and farms (Source 4). The final death toll is thought to be about 20 000 – but many bodies have never been recovered. In Honduras, even optimistic estimates think it will take at least 20 years for the country to repair the damage caused by a disaster which made a million of its 5 million population homeless. The overall cost of damage caused by Mitch is an estimated $10 billion.

Repairing the damage

After Mitch hit Central America, short-term emergency aid in the form of medicines, food, water and shelter came from governments and NGOs across the world.

However, its effects are still being felt today by its peoples, economies and environments. Most of the countries in Central America are relatively poor LEDCs, with economies based primarily on farming. The money needed to repair the damage is simply not available within the region.

Longer term, much of the funding needed to rebuild homes and infrastructure has come from international aid, agencies or organisations like the World Bank. Much of this has been organised via a new Central America Emergency Trust Fund and included money for a road rebuilding project and repairs to schools and clinics. Some of these projects also created jobs for local people. The impact of natural disasters like Hurricane Mitch in LEDCs is far greater than in MEDCs. Long-term recovery is often dependent upon external aid.

| Source 4 | Damage as a result of a mudslide triggered by Mitch | Source 5 | Distributing relief supplies in Honduras after Mitch |

Weather hazards in a MEDC
Hurricane Floyd

Hurricane Floyd hit the east coast of the USA in September 1999. Heavy rain caused flooding across 13 states and led to the evacuation of 4 million people – a million from Florida alone. Over 70 people were killed, the highest death toll in the USA from a hurricane since 1972. The final bill for damage was estimated at $6 billion.

Floyd's progress

Floyd started life on 2 September 1999 as a tropical wave off the West African coast. Five days later it had become a tropical depression, 1500 km east of the Caribbean. A day later it had become a tropical storm. By the time it was 400 km from the Leeward Islands, it had been upgraded to hurricane status. As it turned north-west, the winds started to drop. Turning west once more, it quickly gained strength until winds reached 230 kph and pressure dropped to 921 mbs. Floyd was now a category 4 hurricane, causing widespread damage to the Bahamas as it passed through on 13–14 September.

Mass evacuation in the USA

Still uncertain as to where along the USA coast Floyd would make landfall, the south-eastern states began to evacuate coastal residents inland – Disney World in Florida was shut down for the first time ever. However, Floyd started to move north up the Atlantic coastline, missing Florida. It hit land in North Carolina on 16 September. Wind speeds had dropped as Floyd became a category 2 hurricane. However, it was the rain which did most of the damage, partly because Floyd was a very wide hurricane. Between 16 and 17 September almost 50 cm of rain fell on ground which was still saturated by heavy rain from Hurricane Dennis two weeks earlier.

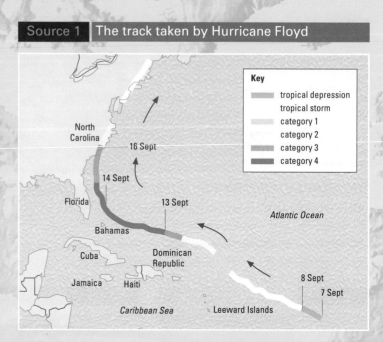

Source 1 — The track taken by Hurricane Floyd

Key
- tropical depression
- tropical storm
- category 1
- category 2
- category 3
- category 4

North Carolina
16 Sept
14 Sept
Florida
13 Sept
Bahamas
Cuba
Dominican Republic
Jamaica   Haiti
Atlantic Ocean
8 Sept
7 Sept
Caribbean Sea   Leeward Islands

Source 2 — Almost 4 million people were evacuated as Floyd approached

Impacts

Storm surges along the coast, up to 3 metres high, and exceptionally heavy rainfall caused extensive flooding across 13 states. All were declared major disaster areas, with North Carolina the worst hit. Rivers peaked at over 7 metres above normal levels, 51 people were drowned and 7000 homes completely destroyed. Tens of thousands homes were damaged and 10 000 inhabitants forced into temporary accommodation. Roads were destroyed and damaged and hundreds of thousands of cattle, hogs (pigs) and poultry were drowned. Electricity supplies were badly affected.

Despite extensive damage to property and land, the quality and timing of early warning systems and subsequent organised evacuation saved many lives. National agencies monitor and track hurricanes via planes and satellites. If a storm is approaching, a 'Hurricane Watch' is announced 36 hours ahead. A Hurricane Warning is issued 24 hours before expected arrival, usually leading to evacuation orders.

Although it took many months before everyone was rehoused and damage repaired, over $2 billion of government aid was made available by Congress. Individual states, insurance companies and business funding also helped recovery. With sophisticated warning systems in place and the ready availability of emergency funds, the impact of tropical storms in wealthy MEDCs like the USA is almost always far less than in poorer LEDCs.

Source 3 Extensive flooding caused widespread damage and over 70 deaths

Source 4 Thousands had to be moved into temporary accommodation like these new trailer homes in North Carolina whilst rebuilding took place

1 Copy the following passage into your book filling in the blank spaces.
The earth's crust is made up of a number of _____ _____. Where these _____ move apart a _____ _____ _____ is formed. Where these _____ push together a _____ _____ _____ is formed.

Beneath the earth's crust is molten rock called _____. This molten rock is brought to the surface by _____ _____. As the molten rock spreads outwards, it pulls the crust apart creating _____ _____. Beneath the sea the molten rock rises to the sea floor producing _____ _____.

Where the moving crust meets another part of the crust, one of them will sink below the other. This junction is called a _____ zone. As it sinks the crust heats up, turns back to molten rock and is forced up as an erupting _____.

2 Make a copy of the volcano (Source 1).
   a Label the following features: magma chamber; main vent; crater; steam; gas and dust; layers of cooled lava.
   b Describe the landforms of intrusive volcanic activity.
   c Explain how and why a volcano erupts.
   d Describe the different shapes of volcanoes and explain how each is formed.

| Source 1 | Volcano |

3 a Using the evidence from the case study about the Kobe earthquake on pages 40–1, make a list of the damage caused.
   b Comparing the information about the smaller Bhuj earthquake in India and the larger Kobe earthquake in Japan, explain why the Bhuj earthquake caused many more deaths than the Kobe earthquake.
   c The main cause of loss of life in an earthquake is the collapse of buildings. Describe the ways in which town planners can reduce earthquake damage to buildings.

4 a Iceland is built by volcanoes. Describe how the people who live on the island use the volcanic resources to their benefit.

b Vesuvius destroyed the city of Pompeii in AD 79. Since then people have settled around the volcano. What are the advantages of living so close to such a dangerous volcano?

5 a Describe the ways in which people try to predict earthquakes and volcanic eruptions.

b Explain why earthquakes and volcanic eruptions are often much more damaging in LEDCs than in MEDCs. Give examples.

6 a Describe how the weather conditions would change as a tropical storm passed over, from the edge through to the eye.

b For either a tropical storm in a LEDC or MEDC, describe the main impacts and damage caused.

c Copy and complete the following table, comparing Hurricane Mitch with Hurricane Floyd.

	Hurricane Mitch	Hurricane Floyd

MEDC or LEDC
Countries affected
Dates
Weather conditions,
    e.g. wind speed, rainfall totals
Effects:
    casualties
    homeless/homes destroyed
    precautions taken
    cost of damage
    other		

UNIT 3

Production

Unit Contents

- Types of employment
- Farming and famine
- Types of farming
- Farming in rich and poor countries
- The changing face of farming: the UK and EU
- Genetically modified crops in a MEDC: North America
- Subsistence rice farming in a LEDC: the Indian subcontinent
- Appropriate technology: farming in Peru
- Large-scale technology: irrigated farming in California, USA
- Intensive market gardening in a MEDC: the Netherlands
- Glasshouse farming in a LEDC: Kenya
- Manufacturing: industrial location
- Manufacturing: industrial change
- High-tech and footloose manufacturing in a MEDC: the M4 corridor in the UK
- Production in a LEDC: India
- World energy resources
- Non-renewable energy
- Renewable energy
- The changing demand for energy in a MEDC: the UK
- Wind power in a MEDC: Cumbria, UK
- Fuelwood in a LEDC: Africa
- Nuclear energy
- Nuclear energy: India

What are the different production methods shown here? Why are they different?

Types of employment

The **production** of goods from **raw materials** and the provision of **services** are essential to our lives. Food production (farming) and the production of manufactured goods are especially important. We use the term 'industry' to describe this range of productive activities. Industries can be classified or grouped according to the types of jobs people do (Source 1): **primary**, **secondary** and **tertiary**. A new, fourth category – **quaternary** – has been added recently.

Source 1 Types of work

Primary industries
These involve the extraction of raw materials to be supplied to other industries.

farming forestry fishing mining

Secondary industries
These are where raw materials are assembled or manufactured to produce finished goods.

food processing car assembly manufacturing building

Tertiary industries
These are jobs which involve providing goods and services for the public.

transport retail medicine catering

Quaternary industries
These include people who provide specialist information and expertise to all the above sectors.

research design engineering computer programming

Employment structure

Employment is the term we usually apply to **formal**, paid work. The percentage of people employed in each type of industry is called **employment structure**. This varies widely between countries and over time. The proportion employed in each sector is a good indicator of the level of development of a country. The pie charts (Source 2) show the employment structures of the UK and Bangladesh. Usually, MEDCs like the UK have a higher proportion of the work-force employed in secondary and tertiary industries. LEDCs like Bangladesh tend to have a higher percentage employed in primary industries like farming.

These proportions often change over time. As a country develops its economy, the proportion employed in primary industry decreases and secondary employment increases. As the economy develops even further, numbers in primary and secondary industries fall, and tertiary becomes the largest employment sector. This is shown in the bar charts for the UK (Source 3).

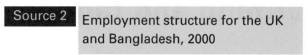

Source 2 | Employment structure for the UK and Bangladesh, 2000

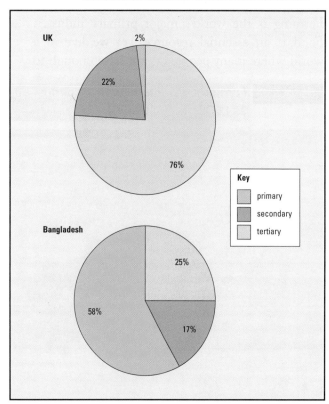

Informal employment

Millions of people across the world, especially in LEDCs, work in the informal sector – sometimes called the 'black economy'. This type of work includes selling goods on the street or providing services, such as shoe-shining. Few informal jobs pay well and most involve long, irregular hours. No taxes are paid and the work is often illegal and often carried out by quite young children. In many places, there is no choice as it is the only way to earn money, and in most cases it contributes significantly to the income of families (see Unit 3.15, pages 80–1).

Source 3 | Changes in employment structure over time, UK

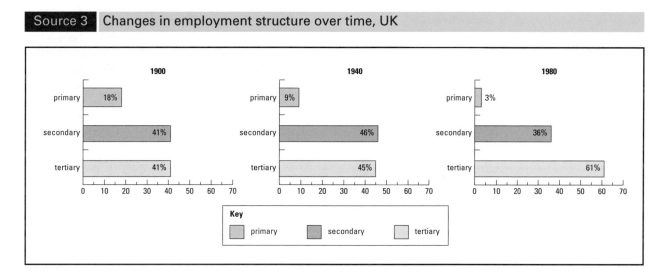

Farming and famine

Farming is the world's major primary industry. Food is an essential resource, yet we live in a world where many people do not have enough to eat. Source 1 shows the parts of the world at risk from **famine**. All the areas shown are in less economically developed countries (LEDCs).

Source 1 Famine areas of the world – almost all countries most at risk are to be found as LEDCs in the tropics

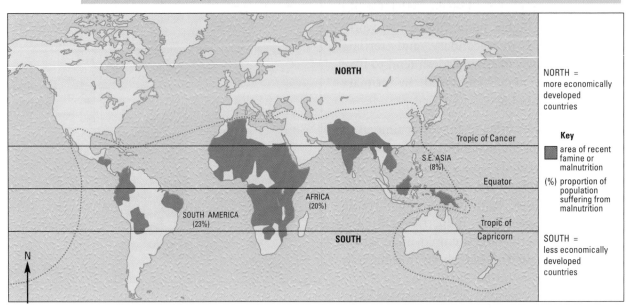

It is estimated that on average an adult needs a balanced diet providing about 2300 calories per day. In the more economically developed countries (MEDCs), calorie consumption is high and the number of overweight people is increasing. This can lead to heart disease and strokes. By contrast, in many African nations people only manage to consume

Source 2 Daily calorie supply

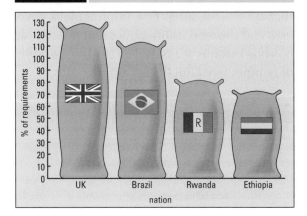

80 per cent of this amount (Source 2). A lack of calories and the right vitamins can cause **malnutrition**, making people weak and sick.

The result of many illnesses is that people become too weak to work. This contributes to the 'circle of hunger' (Source 3), from which it is difficult to escape.

Source 3 The circle of hunger

hunger

tiredness

less to eat

illness

little food is grown

cannot work

The causes of famine

Some people believe that famines are a result of people's laziness or ignorance. The fact is that the causes are complex and can include any one or more of the following:

Drought When the rains fail, harvests can be destroyed and farmers are left without food.

Desertification The removal of trees (deforestation) and overgrazing result in land that is easily eroded and so turns to desert and becomes unproductive.

War Wars can destroy farming as people leave the land to fight or escape, and money is spent on weapons rather than on agriculture. Source 4 tells what happened to one family during the war in Somalia in 1992.

Poverty Landless people do not have land to farm on.

Trade LEDCs get poor prices for the cash crops which they export, yet they pay a high price for manufactured goods which they import from the developed world.

International debt LEDCs owe money to the MEDCs. Many countries have such a huge debt that most of their income goes towards paying off interest on their loans. This leaves little to spend on farming.

What can be done?

Source 5 shows some of the ways in which the MEDCs can help the poorer world. Recently there have been moves by Britain, among others, to 'write

Source 4   Famine as a result of war

Famine in Somalia

The drought affected Abdi Husein's farm near Bardera in Somalia. When it passed he was able to grow corn, tomatoes and olives. But then war reached his village and things got worse.

'I had plenty stored, but they grabbed all my food and took all my animals and personal belongings,' said Husein from his hospital bed.

With his wife, daughter and six sons, Husein made his way to Doblei, the nearest town, eating wild fruits and leaves and gnawing on animal skins. The children died, some from gunfire during the battle, some from hunger.

In Doblei, Husein became more ill, before they heard a rumour that emergency food supplies were on their way.

off' some or all of the international debt, perhaps one of the most important recent moves to help the poor world.

Source 5   'Fairtrade' means a fairer price

We could give aid, but this might only help people in the short term. People might become too dependent on aid in the long term.

Farmers should be given a fair price for their crops. Some companies that do this use the 'Fairtrade' mark. Maya Gold, which is an organic chocolate from Belize, tries to guarantee that peasant farmers are not exploited.

We need to improve the way food is produced. Farmers ought to use more intensive methods such as machinery, fertilisers and pesticides.

Farmers should be given their own land. Small loans would help them buy inexpensive tools, allowing them to use their own knowledge and skills.

Types of farming

Commercial farming is the growing of crops (arable) and rearing of animals (pastoral) for sale at markets. This is the main type of farming in MEDCs.

Subsistence farming is the growing of just sufficient crops and the rearing of just enough animals to feed a family. This is the main type of farming in LEDCs.

Intensive farming uses a small amount of land from which high yields are obtained. One example of this in a MEDC is the kind of market gardening shown in the photograph on page 172. An example from a LEDC is rice growing in India (Source 3).

Extensive farming uses large areas of land from which lower yields are obtained. An example in a MEDC is shown in Source 1 where highly advanced machinery is reaping vast areas of wheat, with hardly a worker in sight. In LEDCs this farming is restricted mainly to animal herding.

| Source 1 | Mechanised wheat farming in Canada – commercial extensive farming |

| Source 3 | Rice cultivation in India – subsistence farmers use little machinery |

| Source 2 | A simple classification of farming |

Type	Commercial	Subsistence
Arable	Extensive cereal cultivation	Intensive rice cultivation to feed family, e.g. India
Pastoral	Ranching of cattle and sheep, e.g. cattle ranching in USA	Herding of cattle and sheep to provide food and hides to feed family or tribe

Farming in rich and poor countries

 3.4

Source 1   Dependence on farming in LEDCs and MEDCs

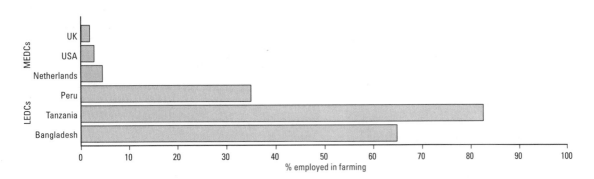

% employed in farming

In LEDCs most people work in the fields to provide enough food for themselves. If there are any crops left over, the smallholders take them to the local markets for sale. Source 2 is a bustling smallholders' market near Arusha in Tanzania.

In contrast, Source 3 shows the more common sight of a superstore selling fresh fruit and vegetables in the UK, a MEDC. Most of this food is produced under intensive commercial conditions in Europe.

Source 2   Duluthi market, Arusha, Tanzania

Source 3   Superstore products in the UK

The changing face of farming: the UK and EU

Source 1 The two pictures below show the same farm in 1963 and in 1991

British farming developed over centuries. However, modern British farming is changing. As technology has improved the physical factors have become less important; more important are the commercial factors. Where to sell the product is a far greater influence than where to grow it.

The typical farm in Britain today employs far fewer workers than in the past and is much more mechanised.

It is now much larger and more specialised than in the past. As mechanisation and the demand for food increased, so farmers were encouraged to join up (amalgamate) their farms. As they did so, many hedgerows were ripped out – a serious loss to the natural wildlife which used them as their habitats.

Farmers were encouraged to sign contracts with major supermarkets and freezing and canning plants. The use of chemical fertilisers and pesticides increased in order to improve yields. This use of chemicals has also had an effect on wildlife and on the water quality of the rivers flowing through the farms.

Modern farming is changing fast. As farming has modernised in the MEDCs the application of technology has produced great improvements, as well as some dangers.

Source 2 Job losses on British farms

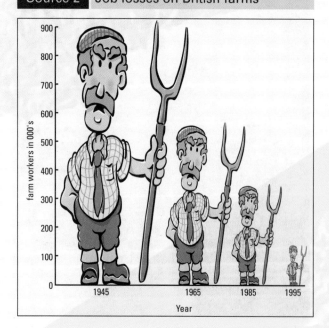

European Union and the Common Agricultural Policy (CAP)

The European Union has encouraged farmers to use their land more intensively. It bought up surplus produce when farmers could not selltheir crops. This led to the creation of hugefood mountains. This food was going to waste and costs of storing it were high, so the CAP was reformed. From 1992 farmers who received grants from the EU had to take up to 20 per cent of their arable land out of production: this is called **set aside**.

The farming industry has been hard hit by these changes and many farmers have had to **diversify**. It is estimated that in 1991 almost 40 per cent of farmers were partly reliant on non-agricultural activities. These include the following:

- **Leisure** Farms with good access to urban areas have found it profitable to open their farms to the public. Some have applied for golf courses to be developed on their land.
- **Tourism** Farms in scenic locations such as the Lake District are increasingly offering bed and breakfast accommodation to tourists.
- **Conifer plantations** Under the Farm Woodland Scheme farmers receive a grant to plant conifers. The growing number of conifers is matched by a huge loss of broad-leaved woodlands.

- **Conservation** Farmers have been encouraged to protect their environment by agreeing to register their land as **Environmentally Sensitive Areas (ESAs)**. Under this scheme farmers receive a payment if they agree to:
  - limit their use of fertilisers
  - restore drystone walls
  - reduce the number of animals they keep.

Beef in crisis

There are fewer cattle on British farms than twenty years ago. As farmers tried to improve the yield of beef and milk, they used feed stuffs containing contaminated products. The result was 'mad cow disease' which was soon recognised as the source of a similar human disease. The effects on British farming were devastating. Foot-and-mouth disease in spring and summer 2001 also caused the slaughter of millions of sheep and cattle.

Organic farming

The beef crisis (Source 3) focused attention on how we produce our food. This helped the already fast growing organic farming industry in the UK. **Organic farmers** use only very small quantities of artificial fertilisers and chemicals on crops. Animals are reared without regular use of antibiotics or other drugs. Crops are not genetically modified (see Unit 3.6, pages 66–7) and farmers work to maintain the landscape and local environment. Whilst costs tend to be higher, there has been a significant increase in the amount of organically produce food being sold, and all the major supermarket chains stock an expanding range of organic produce.

Source 3	Beef in crisis

3.6 Genetically modified crops in a MEDC
North America

The growth and use of **genetically modified** (GM) crops has provoked widespread debate since the first GM products were sold in the early 1990s. Yet growing 'improved' varieties of plants (and animals) has been happening since farming began, with varied success. In the 1950s, scientists discovered that **DNA** carried the genetic detail of living things. By the 1980s, it was possible to identify individual **genes** and transfer them and their specific qualities. This laid the foundation for the world's **biotechnology** companies to develop today's GM industry.

Genetically modified crops are designed to be resistant to competition or destruction by other plants or animals and insects. Genes containing resistance are bred into the new GM crop – but these genes can come from a quite different source, for example fish genes in tomatoes. Some are resistant to herbicides, so competing weeds can be killed with general spray which will not kill the crop. Others produce toxins which kill pests which try to feed off them, reducing the need for pesticides. Animals can also be genetically modified, although there are not yet any GM meat products on sale. Research is taking place on cattle, pigs and fish.

Source 2 shows where GM crops were grown in 2001, covering 52.6 million hectares of farmland (about the size of France). By 2004, 40 countries were growing GM crops either commercially or under trial conditions. Soybeans, oilseed rape and maize (corn) account for the majority of GM food crops, some of which is used for animal fodder. Non-food crops like cotton are also included. Further research is taking place for a variety of rice called 'golden rice' containing high levels of vitamin A.

Source 1	Definitions

Term	Definition
Genetic Modification (GM)	Altering the genetic composition of cells or organisms by introducing 'foreign' genes
genetically modified food	Food product which has some genetically modified organism(s) as an ingredient
DNA	Deoxyribonucleic acid – the genetic base of cells and organisms
gene	A specific sequence of DNA code carrying inherited characteristics
biotechnology	The application of technology to modify the products or processes of living systems

Source 2	Percentage of the world's GM crops grown by country, 2001

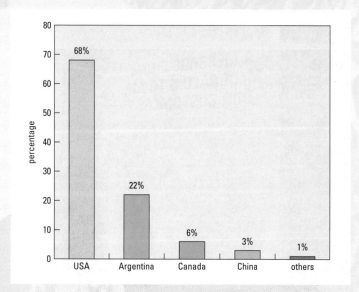

GM in the USA

Sixty-eight per cent of the world's GM crops are grown in the USA with up to 70 per cent of packaged foods on sale in the USA containing GM ingredients. Unlike the fierce debate underway in the EU and other countries (Source 3), there has been little apparent opposition in the USA. This may be in part because such foods have been on sale for over ten years, with no measured effect on human health.

The USA saw the first ever GM food on sale in 1994. Flaver Saver tomatoes were designed to stay fresh longer and resist rotting without altering the taste. Unlike other tomatoes they could be picked and transported when fully ripe. They were withdrawn from sale after two years because consumers were not prepared to pay the higher prices for them.

US company Monsanto is responsible for over 90 per cent of the world's GM crops, and successfully grows soybeans, oilseed rape, maize and cotton. However, its attempts to promote GM wheat have had to be abandoned. The wheat is resistant to the main ingredient in a leading weedkiller called Roundup and also thought to increase yields by over 14 per cent. In 2004 it was forced to abandon plans to grow the wheat commercially. Forty-five per cent of the USA's wheat exports are to Japan and the EU, both of whom are worried about GM wheat cross-pollinating and contaminating the ordinary wheat crop. A threatened boycott worried American farmers, who have pressurised Monsanto to withdraw the plans on commercial grounds.

Source 3	The GM debate: for and against

For	Against
Higher crop yields	Resistance to antibiotics
Cheaper food	Can contaminate ordinary crops
Better quality	Human health effects unknown
Less herbicide/pesticide use	Too expensive for poorer LEDCs

Source 4	GM tomatoes

There is no doubt that the GM debate will continue for many years, both in the USA and across the world. Both sides will need to move from their entrenched positions, and our knowledge will need to be improved and backed by further research for informed decisions to be made.

3.7 Subsistence rice farming in a LEDC
the Indian subcontinent

Source 1 Rice-growing areas of the Indian subcontinent – high temperatures and heavy monsoon rains provide ideal growing conditions

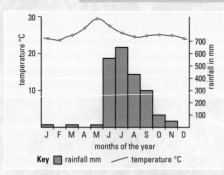

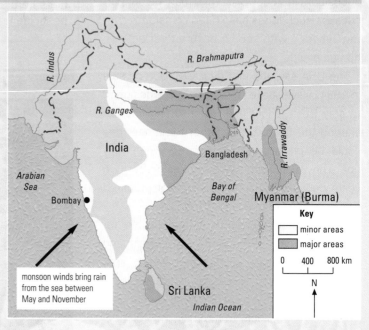

monsoon winds bring rain from the sea between May and November

There can be no doubt that rice is a major crop, given that it feeds one-third of the world's population. The main rice-growing areas of the world are the nations of South-East Asia, including India. Here, rice is grown mainly for subsistence and what little is left is sold.

The main growing areas of India offer the perfect type of climate for rice as temperatures do not fall below 21°C throughout the year (Source 1) and there is a long wet season, called the **monsoon**. The monsoon arrives in May and ends in November. The monsoon winds pick up their moisture over the oceans. The monsoon is then followed by a dry spell which allows the rice to ripen and harvesting to take place.

Rice is the staple crop of most of Asia. Large quantities of water are needed to grow the crop which is usually grown in irrigated fields called **padi fields** (Source 2). Whilst much of this comes from annual monsoon rains, rice farmers have to manage this water so it can be used throughout the year. They build earth banks, or bunds, around their fields and build terraces on sloping land to collect and capture rain when it falls. Some of these flood irrigation systems were built hundreds of years ago. Farmers control how much water they keep or release from each padi or terrace. Water is moved by gravity down slopes or via pumps.

Rice requires a great deal of the farmer's time, especially if more than once rice crop is grown in the same padi over a year. Rice farming is a good example of **intensive** farming.

| Source 2 | A padi field in India |

Many people work in the fields – a labour-intensive farming system

Flooded fields provide ideal growing conditions

Sowing the seeds of change

Traditional varieties of rice meant that there simply was not enough food to go around. As the population of the south-east Asian countries grew rapidly, food could not be grown fast enough. In 1959 the International Rice Research Institute (IRRI) was set up in the Philippines to look at how rice yields could be increased.

Researchers cross-bred two plants: a semi-dwarf plant from China with a strong, tall Indonesian plant. The result was a sturdy, short plant called IR8. How it compared with the traditional variety is shown in Source 3.

| Source 3 | Comparing old and new rice plants |

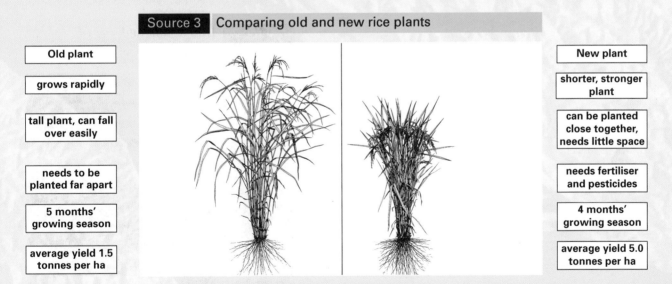

Old plant

grows rapidly

tall plant, can fall over easily

needs to be planted far apart

5 months' growing season

average yield 1.5 tonnes per ha

New plant

shorter, stronger plant

can be planted close together, needs little space

needs fertiliser and pesticides

4 months' growing season

average yield 5.0 tonnes per ha

New plants like IR8 proved to be a success because much more food could be produced. However, they made many demands on farmers. Expensive fertilisers and pesticides were needed and they required much more irrigation. Large-scale irrigation projects meant many small farmers lost their homes. The result has been that the rich have benefited while poor farmers could not afford to grow the new crop.

Plants like IR8 also attracted far more pests than the traditional varieties, and despite the many changes that have been made to IR8, this still remains a problem. The development of GM crops (see Unit 3.6) may revolutionise farming in the LEDCs, producing high-yield plants without the problems of IR8.

Appropriate technology
farming in Peru

Small-scale farming projects can often cause less damage to the environment than large-scale activities and yet be just as effective. **Appropriate technology** uses the skills of local people to find the best ways of improving their farming.

Intermediate Technology is an international development agency which works with people in rural communities in Kenya, Sudan, Zimbabwe, Sri Lanka, Bangladesh and Peru. It works with peasant farmers called *campesinos* in Peru.

Farmers in the highly populated, dry, western area of Peru are dependent on irrigation to grow crops. In recent years poor rains in the sierra mountains have reduced irrigation and water flows, leading to poor harvests. In the Ica Valley (Source 1) this situation is made worse by rapid population growth caused by the migration of people from the drought-stricken sierra. Farmers in the valley depend on irrigation and draw their water either from the River Ica or the Choclocaya Dam in the upper reaches of the river, by way of La Archivana irrigation canal.

However, the way water has been managed favours richer farmers who produce cash crops for export, leaving the *campesinos* with little water for their land.

The aim of Intermediate Technology is to help organise the small farmers so that they can have a fair share of the water:

- farmers now have better access to information about their water rights and are helped to defend their interests through improved organisation
- technical support is also given and the *campesinos* are being helped to rebuild some of their traditional technologies, such as the wooden irrigation gates shown in Source 2.

Source 1 Location of the Ica Valley, Peru

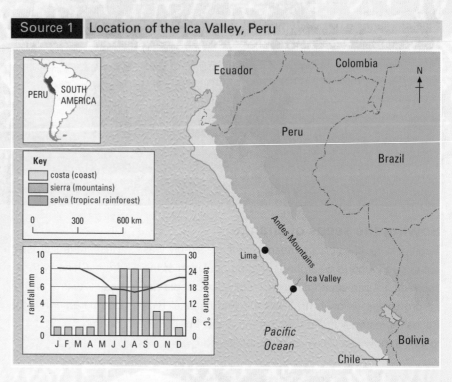

Source 2 Traditional methods of irrigation in the Ica Valley

Primitive well-based irrigation system

Note very dry fields – crops cannot survive without irrigation

Large-scale technology
irrigated farming in California, USA

Irrigation farming of California (MEDC) is based on capital-intensive, large-scale technology.

It is hard to imagine how a desert can be one of the most productive areas of the USA. However, with money and technology the Central Valley desert in California is now a fertile agricultural region.

Much of California has a Mediterranean climate and, as Source 1 shows, this means there is a long summer drought during which little can grow. Farmers in the region faced two specific problems.

1 The Rivers Sacramento and San Joaquin did not supply enough water to **irrigate** the land.
2 Most of the rain fell in the northern part of the valley, yet the most fertile land was in the south.

The Central Valley Project was developed to help distribute water in California more effectively. It is a large irrigation project. Source 1 shows how dams were built in the upper reaches of the main rivers to store water. A series of canals connect these dams (Source 2) to the dry valleys 500 km further south. Water is then brought to the fields by a network of pipelines, sprays and sprinklers.

The cost of the scheme was so high that the land has to be farmed intensively. Only crops which have a high yield and value are grown, such as vines, vegetables and citrus fruits (oranges and lemons). Large-scale intensive commercial farming of this type is called **agribusiness**.

Large schemes like the Central Valley Project are controversial. They have advantages and disadvantages as the table below shows.

Advantages	Disadvantages
Crops can be grown.	Cost – farmers pay a high price for the water. This benefits rich Californian farmers.
Dams can be used for tourism.	
Dams can provide hydro-electric power.	Salts are left behind when water evaporates. This poisons the land.
Industry is attracted to cheap sources of power.	Dams will silt up in time.

Source 1   Irrigation of the San Joaquin Valley

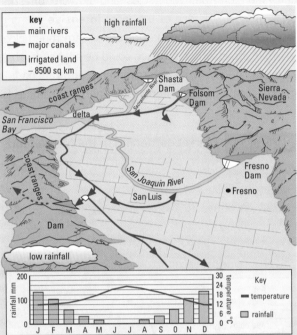

Source 2   The Shasta Dam

Streams from surrounding hills feed the dam

HEP generated from the dam as diverted water turns turbines

The multi-purpose irrigation and power schemes of California are in marked contrast to the small-scale, single-purpose individual projects in Peru.

71

Intensive market gardening in a MEDC
the Netherlands

Market gardening or **horticulture** involves the cultivation of fruit, vegetables and flowers on small plots of land. It is one of the most **intensive** forms of farming. Traditionally market gardens were found close to urban areas where produce could be sold as fresh as possible. Today, with improvements in transport, a market location is less important. The ideal climate for horticulture is a Mediterranean one – a mild winter and an early spring. However, these conditions can be created artificially in glasshouses. The Netherlands has become a specialist region: the coastal area of Westland is called 'the city of glass'.

The Netherlands devotes 27 per cent of its land to farming, with whole areas dedicated to producing fruit, vegetables or flowers (Source 1). There are several reasons for this.

- Almost 50 per cent of the land has been reclaimed from the sea. The new areas of land are called the Polders and have a fertile, peat and clay soil.
- Much of the land is flat, and there are plenty of waterways providing irrigation.
- Dutch farmers have access to cheap natural gas, which is used to heat the glasshouses.
- A good transport network makes it easier to reach the local markets of Amsterdam, Rotterdam and the Hague, and more distant places.
- The European Union has helped the Netherlands to expand its market throughout Western Europe.

What makes market gardening intensive? One farmer in Westland, south of the Hague, described what made his farm intensive. 'The Polder lands are very expensive, so the land has to be used almost continuously, using crop rotation, to keep the soil fertile. My farm is 2.25 hectares, which is more than twice the average size. Market gardening is labour-intensive, so we need plenty of people to work on the farms especially at harvest time. It is a costly business, and we can only sell produce which will give us a high price, if we are to make a profit.'

| Source 1 | Market gardening in the Netherlands |

Key
- vegetables
- flowers, shrubs and bulbs
- fruit
- arable
- pasture
- reclaimed land
- arable on reclaimed land

0   60 km

North Sea · Haarlem · Amsterdam · Utrecht · The Hague · Westland · Rotterdam · The Netherlands · Belgium

| Source 2 | Intensive market gardening produces a pattern of colour |

Glasshouse farming in a LEDC
Kenya

Horticulture in Kenya has become one of the country's most important export industries, accounting for two-thirds of total agricultural exports. Half a million people depend on the industry and the 135 000 workers who grow, cut and package fruit, vegetables and flowers for export. Most of the flower farms are found around Lake Naivasha, north-west of Nairobi.

Kenya now supplies more flowers to the world market than any other LEDC except Colombia. Flower growing began to develop in the 1970s when a Danish company, DCK, was attracted to Kenya by incentives from both the Danish and Kenyan governments. The project lasted about ten years before Brooke Bond (the tea company) and then Sulmac took over some of the farms. Today, Sulmac and ODC (a Dutch company) are the industry leaders in Kenya.

Seventy-five per cent of the flowers are grown on 30 large farms around Lake Naivasha and transported by air to markets in Europe. Soils around the lake are fertile and the lake provides water for irrigation. With intermittent droughts over the past five years, pressure on the lake has increased. The population of the area has risen from 50 000 to 350 000 in 20 years, but services have not kept pace with this, resulting in untreated sewage being discharged into the lake. Fertilisers and pesticides also run off and pollute the lake. Water levels have dropped and local fisherman have seen a decrease in catches. Deforestation is also a threat as trees are cut down for fuelwood.

The flower industry in Kenya has its own regulatory body, the Kenya Flower Council (KFC). They have introduced a strict code of practice for their members, encouraging recycling of water and using safe pesticides.

Source 1 Lake Naivasha, Kenya's main flower growing region

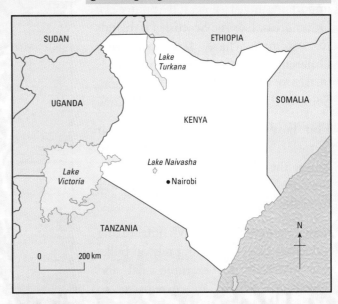

Source 2 Kenya's thriving flower industry

Manufacturing: industrial location

Economic activities are not evenly spread in most countries. Some areas have high concentrations of industry – for example, in the UK the Midlands remains an important industrial region (Source 1). Other areas, like the Scottish Highlands for example, have few industries. Different industries have specific needs or factors which influence their location. Some of these factors are outlined below.

Flat land Some industries, like car factories, need large areas of flat land. Often the cheapest and most suitable sites are on **greenfield locations** away from the city.

Raw materials Manufacturing industries like steelmaking rely on bulky raw materials, for example coal and limestone, which are expensive to transport. As a result many traditional **heavy industries** are located close to their raw materials.

Energy In the past it was important for factories to be close to power supplies, for example textile mills were built close to supplies of fast-flowing water. Today the widespread availability of electricity makes this less important.

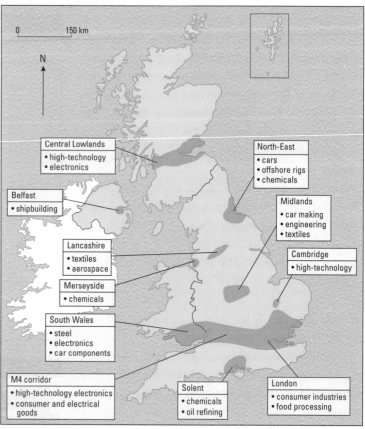

Source 1 | Industrial areas in the UK

0   150 km

N

Central Lowlands
• high-technology
• electronics

North-East
• cars
• offshore rigs
• chemicals

Belfast
• shipbuilding

Midlands
• car making
• engineering
• textiles

Lancashire
• textiles
• aerospace

Cambridge
• high-technology

Merseyside
• chemicals

South Wales
• steel
• electronics
• car components

M4 corridor
• high-technology electronics
• consumer and electrical goods

Solent
• chemicals
• oil refining

London
• consumer industries
• food processing

Labour force Industries that rely on a large workforce, like car production, tend to be found close to or within easy reach of cities where many of their workers and customers live. Source 2 shows another example of a **labour-intensive** factory: clothing factories are often found in inner cities.

Markets Manufacturers do not like to be far from their markets as this increases costs. One reason why Sony decided to make televisions in the UK was to be nearer to its European markets. The EU is a market of nearly 350 million, mainly wealthy, people.

Transport links Factories need to be located close to good transport links to ensure that the raw materials they need and the finished products they manufacture are moved with ease. Many prefer to locate near to motorway junctions.

Source 2 | An inner city clothing factory

Source 3 Assisted regions from 1996

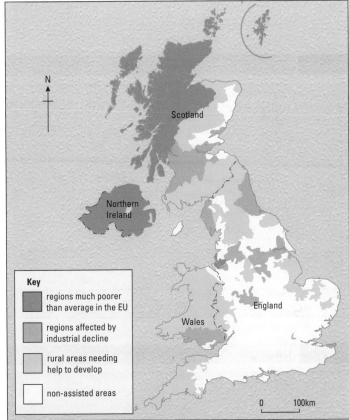

Key

- regions much poorer than average in the EU
- regions affected by industrial decline
- rural areas needing help to develop
- non-assisted areas

Scotland

Northern Ireland

England

Wales

0   100km

Source 4 Industrial estate

close to city for workers

purpose-built buildings

car parks

cheap land

close to roads

What can governments do?

Where regions have lost industries or unemployment is high, then governments may provide money or other forms of help to attract new investment. For example, the EU has a regional policy to promote new development in its less wealthy regions (Source 3). Help may come in the form of:

- giving businesses rent-free periods, grants and loans
- infrastructure: building new roads, water supplies and electricity
- retraining workers – to provide labour with relevant skills
- providing sites for business parks and new start-ups.

New jobs in new places

Some secondary activities can be grouped according to their location.

Heavy industries rely on bulky raw materials and tend to be found close to them, to reduce transport costs.

Light or **footloose industries** can locate almost anywhere, provided that communications are good. In the past, manufacturing industries were found close to their raw materials. Now, with developments in transport, industries do not have to be tied to a certain location. They can be footloose. Today many products are made on industrial estates (Source 4) located close to cities. Offices are attracted to business parks, usually in out-of-town locations.

Highly specialised quaternary activities like research take place in **science parks**.

The widespread use of computers, modems and the Internet will probably allow more people to work from home in the future or in small communities known as **tele-cottages**, such as those already in existence in rural areas of South Wales.

75

Manufacturing: industrial change

Workplaces rarely stay the same. The way goods and services are produced and delivered has changed dramatically over the last century. This has had a huge impact on people's lives and on the location of work in many countries, for example the UK.

One of the biggest changes has been the loss of jobs in the primary and secondary sectors (Source 1). Since the Second World War (1939–45) there has been a decline in the number of people working in mining and manufacturing in the UK. Reasons for this include the following:

- Mechanisation and **automation** meant that fewer people were needed in factories.
- Many firms did not spend enough money on updating their factories.
- Competition from certain newly industrialised countries like South Korea and Taiwan, who can manufacture products more cheaply and efficiently.

As thousands of jobs disappeared it was the regions of traditional heavy industry, such as the North-East, that suffered the most, as Source 2 shows.

The impact of change

Unemployment can have a devastating effect upon individuals, families and society at large. It can result in:

- high levels of poverty
- increased levels of crime and vandalism
- low educational achievement and expectations.

| Source 1 | Change in UK employment |

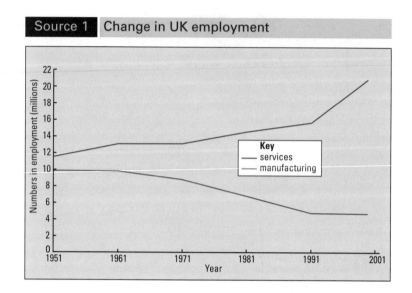

| Source 2 | Unemployment by UK region in 2000 |

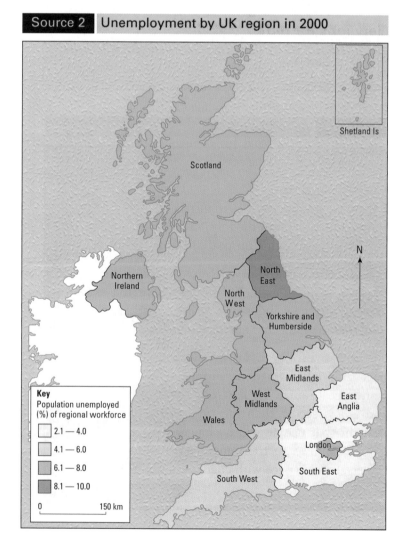

Other effects of industrial decline are that:

• people may lose their homes
• local shops close because people can't afford to use them
• families may break up.

Service industries

Not all areas of industry have been in decline over the last 100 years, for example many new jobs have been created in **service industries**. Service industries include activities as diverse as tourism, banking and transport. Much of the growth has been concentrated in urban areas, especially in southern Britain. Here, the importance of London as a financial centre has played an important role in attracting new jobs. Economic decline in the north and west and growth in the south have led to **regional inequalities**. However, in the late 1980s the technological revolution began to affect service industries and many jobs have been lost in sectors like banking as cash machines have replaced bank clerks (Source 4).

New jobs for old?

Areas affected by industrial decline were given EU and government assistance (see page 79) to attract new industry. This has led to a steady stream of inward investment by companies from the Far East like Sony and Toyota. The head of a Taiwanese firm opening a new TV factory in Scotland said his company was attracted to Britain for the following reasons:

• low labour costs
• there are few strikes – so industrial relations are good
• government help in the form of grants
• to be close to their European markets.

One drawback of the new jobs created is that many of the long-term unemployed do not have the skills to work in the new industries. While some people have been retrained, this has not been possible for all workers.

There has also been an increase in the number of **part-time jobs** which has attracted women into the workforce (Source 5). Part-time work often pays less and is less secure than full-time work. In future it is thought that people will no longer have a job for life. They will need to train and retrain throughout their careers, to keep pace with changes in the workplace.

Source 3 A new service industry – the call centre: this one is located on an out-of-town business park

Source 4 Machines like this cost jobs

Source 5 Women play an active role in the workforce

High-tech and footloose manufacturing in a MEDC
the M4 corridor in the UK

The M4 corridor is a zone of mainly footloose high-tech industries. It stretches from London westwards to Bristol, following the route of the M4 motorway. The greatest concentration of industries is in the section between London and Reading, but towns further west such as Swindon and Bristol have a long history of production. There are many food and drink companies. Electrical and household goods are assembled and packed for distribution. Honda cars are made in Swindon. Over the past twenty years many high-tech companies have arrived. Oracle and NEC in Reading are just two examples. Many are engaged in research and development in the quaternary sector, as well as making telecommunications equipment, micro-electronics and computers.

Source 1 The M4 corridor

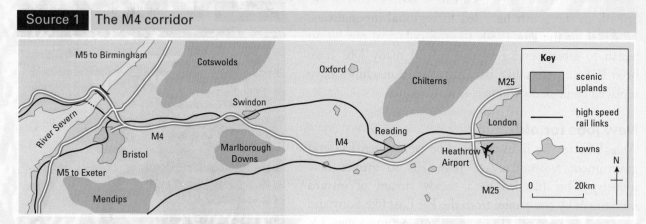

Footloose industries are mainly light or high-tech industries and have considerable freedom in location. This means that most have chosen to locate in the M4 corridor. They have studied its advantages and decided it is the best place for their company to be located. What attracts industries to the M4 corridor? The favourable factors can be arranged under three main headings.

- **Transport:** This is the most important factor for many industries. The M4 links into the UK's other major motorways allowing easy assembly of raw materials and distribution of finished products. The high speed rail link from London to South Wales runs through the middle of the corridor. Heathrow Airport lies between Reading and London allowing international contacts.
- **Site:** Many high-tech industries and science parks are located close to motorways. Not only does this speed up distribution, it gives companies a wider choice of location or sites away from congested urban areas. Land in such areas is usually much cheaper and in attractive undeveloped areas.
- **Market:** The wealthiest market in the country is concentrated in London and the South-East. Motorways give access to markets elsewhere in the UK. To the east of London there are motorway and Eurostar links to the Channel Tunnel and the rest of the EU.

Source 2 Examples of companies which have chosen to locate in the M4 corridor

- **Labour:** High-tech companies in particular need skilled scientists and engineers. There are many places of research in the M4 corridor producing trained people. These include universities in Bristol, Oxford and Reading. These people are often happy to live in this region because of the nearness of London and its facilities, and because the corridor is surrounded by some scenic uplands such as the South Downs and Cotswolds. There is good countryside for recreation at weekends and in holidays.

Source 3   Imperial Park, South Wales

- **excellent road and rail links**
- **spacious parkland setting**
- **9140 m² office space available**
- **close ties with Imperial College of Science and Technology**
- **future links with Cardiff University planned**
- **purpose-built accommodation**
- **support from Newport Borough Council and the Welsh Development Agency**

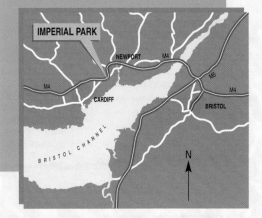

A science park: Imperial Park, Newport

In Newport in South Wales, the Welsh Development Agency together with Newport Council have been keen to encourage the development of high-tech industries. They achieved this by creating a science park named Imperial Park. This opened in 1994. Source 3 shows what it offers.

Here, good access, a scenic location and links with local universities are important locational factors. By 1996, there were eight companies operating in Imperial Park, specialising in food testing and software design. Imperial Park prides itself on being an extension of the M4 corridor, which is home to many science parks and high-tech industries.

Production in a LEDC
India

Despite a population of 1 billion people and a wealth of natural resources, India remains one of the world's poorest nations.

Source 1 compares India with the UK. Despite its poverty, the Indian government has invested vast amounts of money trying to **industrialise** the country.

Planning for growth

After independence from Britain in 1947 India put into action a series of five-year plans. These were based on a communist model of development, called **import substitution**. This is where a country tries to produce all the goods it needs so it does not have to import products from other countries. Source 2 summarises what the plans tried to achieve.

Source 1	India compared with the UK	
	UK	India
Area (in km²)	241 595	3 287 590
Population (in millions)	59.2	1000
Population density (per km²)	241	280
Natural increase (%)	0.2	1.7
Life expectancy (years)	76	61
GNP per capita (US $)	18 620	1348

Source 2 India's five-year plans

Plan	1–3	4	5–9
Years	1951–66	1969–74	1979–1990
Aims	Develop heavy industry	Develop rural area	Develop infrastructure
Example	Damodar Valley natural resources like coal, ore and bauxite used to develop heavy industry	the green revolution investment in high yielding seeds	increase power supplies improve irrigation new roads and railways

Source 3 India's main industrial regions

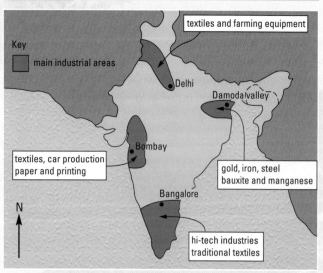

textiles and farming equipment

Key
main industrial areas

Delhi

Damodar valley

Bombay

textiles, car production paper and printing

gold, iron, steel bauxite and manganese

Bangalore

hi-tech industries traditional textiles

N

A great deal of money was spent on developing manufacturing industries producing a range of goods. Source 3 shows some of the main regions which emerged. While heavy industry was concentrated in the north-east in the Damodar Valley, newer high-tech industries were being encouraged in cities like Bangalore further south. Despite these plans, Indian industries continued to face a number of problems.

- Many areas suffered power shortages.
- India continued to import more than it exported – this led to **trade deficits**. India especially relied on oil. When prices rose rapidly in the 1970s this had a serious effect on Indian industry.
- Poor transport links made it hard to develop industry in more remote regions.
- Corruption and inefficiency led to waste.

For almost 50 years the Indian government supported and developed large-scale manufacturing. However, smaller, privately run firms are found across India, producing goods for a local market, but many of them are limited in what they can sell because people are too poor to afford consumer goods.

The informal sector

Unemployment is a huge problem in India, so people will turn their hand to a variety of activities in order to make a living. These might range from shoeshining to selling scrap metal. These are all examples of work in the informal sector. This type of work is characterised by insecurity, no taxes are paid and often workers receive 'cash in hand' payment. It is difficult to estimate how many people work in this sector because informal activities are illegal (Source 4).

A new future

Since the early 1990s the Indian government's approach to production has changed. It is now keen to encourage foreign companies to open factories in India. It especially wants to develop high-tech industries. One way forward is to create technology or science parks (Source 5).

The Indian economy is growing rapidly by more than 6 per cent per year. There is great potential for further growth because of the size of the home market. India is the world's second most populous country. However, major transport problems have yet to be solved. Indian industrialists regularly complain that their businesses are being held back by the need to move goods along pot-holed and congested roads. The passage of goods through overcrowded ports and airports is slow. The price of moving a container 1200 km by road from New Delhi to Bombay costs half as much again as to send it by sea to Europe. Jams on the information super-highway are frequent; using it at the best of times is not helped by the erratic nature of much of the country's electricity supply.

Source 4	Children often work in the informal sector

Source 5	Information Technology Park, Bangalore

The main features of the Information Technology Park are the following.

- Space for offices, shops, luxury homes and parks.
- It is 18 km east of Bangalore and 20 minutes away from the airport.
- State-of-the-art modern buildings, power supply and communications.
- It is designed for technology-oriented companies.
- A landscaped park-like environment.
- It is India's first science park.

World energy resources

Energy is one of the most important of all the world's resources. We need energy to keep us warm and to cook with. It gives us light and drives machinery for transport and industry. Fortunately our natural environment provides us with a wide range of energy sources. **Fossil fuels** are coal, oil and natural gas. Fuelwood, uranium, flowing water, the wind and the sun can be used to produce energy. Source 1 shows the relative importance of each of these in 2000.

Energy sources such as fossil fuels are classed as **non-renewable** – once used up they cannot be replaced. Newer energy sources are often **renewable**, for example solar and wind power – they can be used again and again. These are often **sustainable**, and are likely to play an increasingly important role in the future.

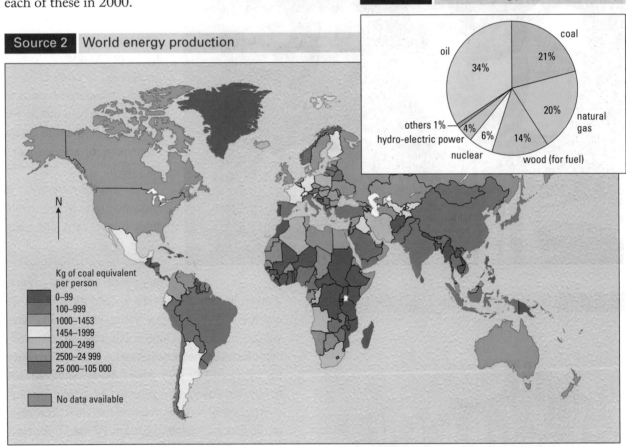

Source 2 | World energy production

Source 1 | World energy sources

Who supplies the world's energy?

Seventy-five per cent of the world's energy is produced by burning fossil fuels. Source 2 shows that these are not very evenly distributed across the world. The USA, Canada, former USSR, Western Europe, Australia, China and the Middle East contain most of the world's coal, oil and natural gas. Western Europe and North America were the first regions to become industrialised. Their early industrial development was helped by the abundance of fossil fuels, particularly coal. Many of the world's MEDCs are found here today.

Source 2 does not take into account the use of wood and **biomass** fuels (fuels made from burning or rotting plants and vegetation). Many people in LEDCs rely on these to supply their energy needs.

Who uses the world's energy?

Europe and North America use 70 per cent of the world's energy (Source 3), although only 20 per cent of the world's population live there. These regions developed their industries quickly using fossil fuels. Today, with many of their own reserves falling or exhausted, they need to import energy to meet demands, especially oil.

Source 3	World energy consumption

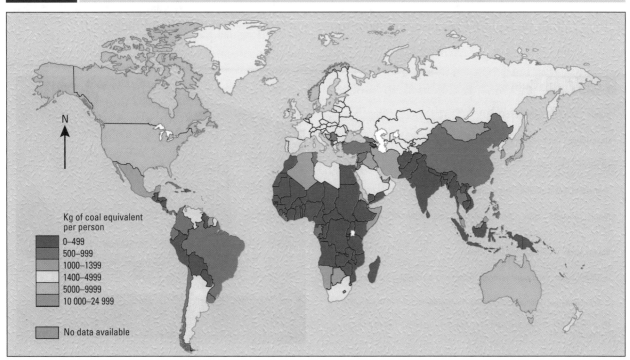

Source 4	Energy in the city

A comparison of Sources 2 and 3 shows quite clearly that the world's major producers of energy are also the major consumers. The amount of energy a country uses is a good indicator of its stage of development (Source 4). The presence of energy resources has obviously been a major factor in the industrial development of some countries. Many of today's MEDCs have substantial fossil fuel deposits – or have had them in the past. The main exception is Japan, whose industry has developed despite a lack of energy resources. Even so, Japan is looking to nuclear power in the future, rather than continuing to rely on importing oil.

As fossil fuels start to run out and countries become more aware of the environmental problems caused by their use, the relative importance of different types of energy seems likely to change.

Non-renewable energy

The world's energy resources can be divided into non-renewable and renewable resources. Non-renewable resources are finite – once they are used up they cannot be replaced because they take too long to form or regrow. They include the major fossil fuels formed over tens of thousands of years – coal, oil and natural gas, plus uranium (used in nuclear power stations) and fuelwood.

Fact File | Coal

Status   non-renewable fossil fuel

Description   formed underground from decaying plant matter

Lifespan   230 years

% share of world energy consumption   21

Main producers   USA, China, Australia, India, South Africa, Russian Federation

Energy uses   electricity, heating, coke

✓ advantages   high world reserves; newer mines are highly mechanised

X disadvantages   pollution – CO_2, the major greenhouse gas responsible for global warming: SO_2, the main gas responsible for acid rain; mining can be difficult and dangerous; opencast pits destroy land; heavy/bulky to transport

Fact File | Oil

Status   non-renewable fossil fuel

Description   formed underground from decaying animal/plant matter

Lifespan   41 years

% share of world energy consumption   34

Main producers   Saudi Arabia, USA, Russian Federation, Iran, Mexico, Venezuela, China

Energy uses   electricity, petroleum, diesel, fuel oils, liquid petroleum gas, coke and many non-energy uses, e.g. plastics, medicines, fertilisers

✓ advantages   variety of uses; fairly easy to transport; efficient; less pollution than coal

X disadvantages   low reserves; some air pollution; danger of spills (especially at sea) and explosions

Fact File | Natural gas

Status   non-renewable fossil fuel

Description   formed underground from decaying animal/plant matter; often found with oil

Lifespan   62 years

% share of world energy consumption   20

Main producers   Russian Federation, USA, Canada, UK, Algeria

Energy uses   electricity, cooking, heating

✓ **advantages**   efficient; clean – least polluting of the fossil fuels; easy to transport

X **disadvantages**   explosions; some air pollution

Fact File | Fuelwood

Status   non-renewable fossil fuel

Description   trees, usually in natural environment, not grown specifically for fuel

Lifespan   variable within each country, but declining

% share of world energy consumption   14

Main producers of energy   LEDCs, especially in Africa and Asia

Energy uses   heating, cooking (also used for building homes and fences)

✓ **advantages**   easily available, collected daily by local people; free; replanting possible

X **disadvantages**   trees quickly used; time-consuming – wood collected daily; deforestation leading to other problems (soil erosion, desertification); replanting cannot keep pace with consumption

Fact File | Nuclear

Status   non-renewable

Description   heavy metal (uranium) element found naturally in rock deposits

Lifespan   unknown

% share of world energy consumption   6

Main producers of energy   USA, France, Japan, Germany, Russian Federation

Energy uses   used in a chain reaction to produce heat for electricity

✓ **advantages**   clean; fewer greenhouse gases; efficient; uses very small amounts of raw materials; small amounts of waste

X **disadvantages**   dangers of radiation; high cost of building and decommissioning power stations; problems over disposal of waste; accident at Chernobyl raised public fears; Sellafield (Cumbria) has had a number of minor leaks

Renewable energy

Fossil fuels are non-renewable energy sources. However, there are many sources which can be classed as renewable sources. These include the use of water – hydro-electric power, tidal and wave; the wind; the sun; geothermal and biomass/biogas.

Renewable resources are generally cleaner than non-renewable sources, but as yet produce only 6 per cent of the world's energy needs. Solar, tidal, wave, geothermal and biomass/biogas are often called 'alternative' energies.

Fact File   Hydro-electric power

Status   renewable

Description   good, regular supply of water needed; water held in a reservoir, channelled through pipes to a turbine

% share of world energy consumption   4

Main producers   Canada, USA, Brazil, China, Russian Federation

Energy uses   electricity

✓ advantages   very clean; reservoirs/dams can also control flooding/provide water in times of shortage; often in remote, mountainous, sparsely populated areas

X disadvantages   large areas of land flooded; silt trapped behind dam; lake silts up; visual pollution from pylons and dam

Fact File   Tidal

Status   renewable

Description   tidal water drives turbines

% share of world energy consumption   insignificant

Main producers   France, Russian Federation

Energy uses   electricity

✓ advantages   large schemes could produce a lot of electricity; clean; barrage can protect coasts from erosion

X disadvantages   very expensive to build; few suitable sites; disrupts coastal ecosystems and shipping

Fact File — Solar

Status renewable

Description solar panels or photovoltaic cells using sunlight

% share of world energy consumption less than 1

Main producers USA, India

Energy uses direct heating, electricity

✓ **advantages** could be used in most parts of the world – unlimited supplies; clean; can be built in to new buildings; efficient

✗ **disadvantages** expensive: needs sunlight – cloud/night = no energy; large amounts of energy require technological development and reduction in costs of PVs (photovoltaic cells)

Fact File — Wind

Status renewable

Description wind drives blades to turn turbines

% share of world energy consumption less than 1

Main producers Denmark, California USA

Energy uses electricity

✓ **advantages** very clean; no air pollution; small-scale and large-scale schemes possible; cheap to run

✗ **disadvantages** winds are unpredictable and not constant; visual and noise pollution in quiet, rural areas; many turbines needed to produce sufficient energy

Fact File — Geothermal

Status renewable

Description boreholes can be drilled below ground to use the earth's natural heat; cold water is pumped down, hot water/steam channelled back

% share of world energy consumption less than 1

Main producers Japan, New Zealand, Russian Federation, Iceland, Hungary

Energy uses electricity, direct heating

✓ **advantages** many potential sites, but most are in volcanic areas at the moment

✗ **disadvantages** sulphuric gases; expensive to develop; very high temperature can create maintenance problems

Fact File — Biogas/biomass

Status renewable

Description fermented animal or plant waste or crops (e.g. sugar cane); refuse incineration

% share of world energy consumption less than 1

Main producers Brazil, Japan, Germany, Denmark, India

Energy uses ethanol, methane, electricity, heating

✓ **advantages** widely available, especially in LEDCs; uses waste products; can be used at a local level

✗ **disadvantages** can be expensive to set up; waste cannot be used in other ways, e.g. fertilisers; some pollution

The changing demand for energy in a MEDC the UK

The 1980s and 1990s saw great changes in the UK's energy demand. The greatest changes were in the electricity industry. No longer is the UK so heavily dependent on electricity generated in coal-fired power stations, like the one shown in Source 1. Source 2 shows that many of the country's coal-burning power stations are located in the Midlands and the North, close to the coalfields.

Source 1 Coal-fired power station

Source 3 UK energy generation

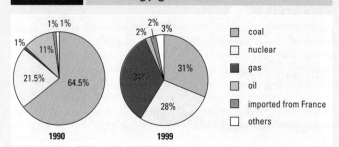

1990: 1% 1%, 1%, 11%, 21.5%, 64.5%
1999: 2% 3%, 2%, 34%, 31%, 28%

Key: coal, nuclear, gas, oil, imported from France, others

Source 2 UK power stations

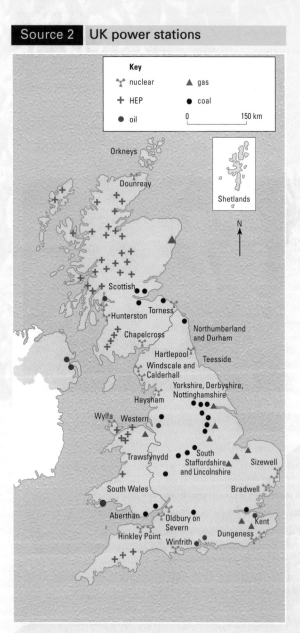

Key
- ⚛ nuclear
- ➕ HEP
- ● oil
- ▲ gas
- ● coal

0   150 km

Orkneys
Shetlands
N
Dounreay
Scottish
Hunterston
Torness
Chapelcross
Northumberland and Durham
Hartlepool
Windscale and Calderhall
Teesside
Yorkshire, Derbyshire, Nottinghamshire
Heysham
Wylfa
Western
Trawsfynydd
South Staffordshire and Lincolnshire
Sizewell
South Wales
Bradwell
Aberthan
Oldbury on Severn
Kent
Hinkley Point
Winfrith
Dungeness

However, Source 3 shows how much the pattern of energy sources used to generate electricity in the UK has changed in ten years. Coal has suffered a great decline in relative importance, whereas natural gas has seen a massive increase in its importance. This has been called the 'dash for gas' to generate electricity. Nuclear power has risen as well, but by a much smaller percentage. Included in the 'others' category are renewable energy sources such as HEP and wind power. As you can see from Source 3, these still contribute only a small amount to the total electricity output of the UK, although future contributions are expected to be greater.

Source 4 Decline in coal in the UK

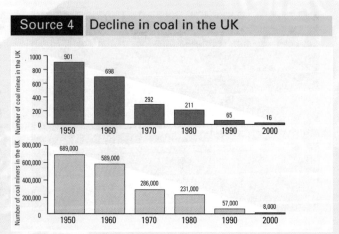

Number of coal mines in the UK
- 1950: 901
- 1960: 698
- 1970: 292
- 1980: 211
- 1990: 65
- 2000: 16

Number of coal miners in the UK
- 1950: 689,000
- 1960: 589,000
- 1970: 286,000
- 1980: 231,000
- 1990: 57,000
- 2000: 8,000

Decline in coal

Source 4 shows just how fast and dramatic the decline in coal mines and miners has been in the UK since 1950. There are a number of reasons for this decline. In many mines the seams of coal were narrow and expensive to work. In more modern mines, where coal seams were thicker, mechanisation meant that fewer men were needed to mine the same amount of coal. In many years it was cheaper to import coal from overseas. Burning coal releases more gases into the atmosphere than the other fossil fuels. Coal-fired power stations are a major cause of acid rain. Coal is dirty, bulky and expensive to transport compared with oil and gas. Only 16 deep mines were still working in the UK in 2000. Closing the mines caused many problems in the mining villages and towns. The number one problem was unemployment.

The dash for gas

Natural gas is a cleaner fuel than coal. When it burns, much less carbon dioxide is emitted. The UK has its own gas fields in the North Sea and off the west coast, for example in Morecambe Bay. In fact, the UK is the world's fourth largest producer of natural gas. Once a gas field is in production, little labour is needed and gas is much cheaper to obtain than coal. It is easy to pipe the gas onshore and then distribute it cheaply through the network of pipelines to homes, factories and power stations. Governments have encouraged electricity companies to change to gas to generate electricity, because it is helping them to meet their targets for reduced emissions of carbon dioxide into the atmosphere.

Little change in nuclear power

Sizewell B is the UK's only pressurised water reactor. It cost £2 billion to build and was opened in 1975 after many years of delay. There is now so much opposition to nuclear power from environmental groups and from the general public that the old Magnox reactors, which are coming to the end of their useful lives, will not be replaced. Therefore the percentage of electricity from nuclear power stations will soon start to go down.

Renewable energy – the future?

Concerns over **global warming** and **greenhouse gas** emission, plus the increasing demand for energy has forced governments like those in the UK to plan to reduce their reliance on non-renewable fossil fuels. Targets are now in place to produce 10 per cent of the UK's electricity from renewable sources by the year 2010. In the UK this is likely to result in an increase in the number of wind farms – both on land (see page 90) and offshore. Other likely sources are an increase in hydro-electric power and the use of waste to produce **biogas**.

| Source 5 | Sizewell B nuclear power station in Suffolk |

3.20 | Wind power in a MEDC
Cumbria, UK

At present less than 1% of the UK's energy is produced by alternative sources, such as wind and solar power. Recent government legislation requires the industry to develop these sources through NFFO (Non Fossil Fuel Obligations).

Sixty wind projects have been approved in the UK. Electricity is generated by two or three-bladed **turbines**, usually built in groups of ten to a hundred creating **wind farms**. An average wind speed of at least 5 metres per second is needed. The map (Source 1) shows where such conditions are found in the UK. In future up to 10 per cent of the UK's electricity could be generated this way.

Ideal sites for wind farms are often in rural areas. Wind power is clean, but electricity cannot be generated when the wind stops. People are concerned about the noise of the turbines and the visual pollution spoiling large areas of countryside. In 2000 the first two offshore wind turbines started to produce electricity in the North Sea near Blyth in Northumberland.

Haverigg Wind Farm, Cumbria

The Haverigg Wind Farm (Source 2) is built on a disused airfield at a cost of £1 million. Five separate three-bladed wind turbines have been built facing the coast, using the prevailing south-westerly winds. The turbines are each 120 m apart and 30 m high. Computers control each turbine, turning the blades to catch the wind.

The wind farm took two months to build – access roads were not needed but concrete bases were needed for the turbines. Haverigg began electricity generation in August 1993.

Source 1 Wind farms and possible sites

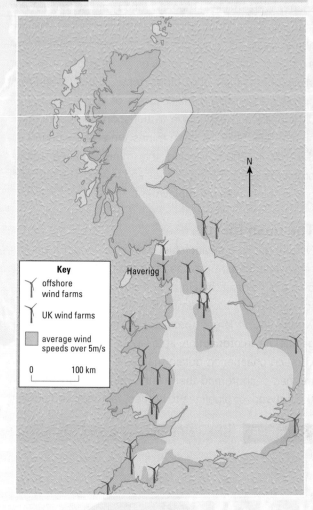

Key
- offshore wind farms
- UK wind farms
- average wind speeds over 5m/s

0   100 km

Haverigg

N

The benefits of Haverigg Wind Farm are as follows:

- the site is a disused airfield; turbines are unlikely to cause objections on the grounds of noise or visual pollution
- a full EIA (Environmental Impact Assessment) was carried out; planning authorities and local people were consulted
- enough electricity is generated per year (3 Gwh) to meet the needs of 500 homes
- local firms were used for construction work, bringing income to the area
- the surrounding farmland can remain in use
- the electricity produced is clean and renewable.

Source 2 Haverigg Wind Farm

Fuelwood in a LEDC
Africa

Whilst the world's MEDCs are looking to develop alternative energy sources to replace oil, coal and gas, many of the world's LEDCs are suffering from a fuel crisis of their own. Fifty per cent of the world's population use wood as their only source of fuel for cooking and heating (Source 1).

The **fuelwood** crisis is especially acute in African countries like Niger and Burkina Faso, south of the Sahara desert in a region known as the Sahel (Source 2). As the population has grown, so has the need to use more wood for cooking. Wood is also cut down to create farmland. As a result, people have to walk further and further from their homes to collect fuelwood as trees close to home have been cut down (Source 3). Trees are being used at such a rate that even replanting programmes cannot keep pace. People often cannot find enough wood and have to find money to buy it. As it becomes even more scarce, prices get higher.

| Source 1 | Fuelwood is gathered on a daily basis |

What can be done?

- More efficient ways of using and managing existing resources are necessary. Improved cooking stoves or ovens, not open fires, would use far less fuel and lose far less heat.
- Woodland needs to be managed carefully, with new trees replanted to replace those used up.
- If wood is an essential fuel, its other uses, e.g. fencing and building, need to be met by using alternative materials, e.g. wire, bricks etc. Once woodland is removed, bare soil may be lost through erosion and land becomes unproductive.

| Source 2 | The Sahel region |

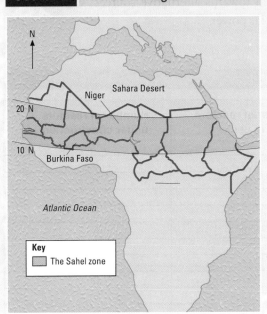

N

Sahara Desert

Niger

20 N

10 N

Burkina Faso

Atlantic Ocean

Key
The Sahel zone

| Source 3 | Replanting fuelwood cannot keep pace with consumption |

Journey time increases as trees are cut down

1/2 day journey

1 hour journey

In Niger a manual labourer now has to spend one quarter of the family income on wood

1 day journey

Nuclear energy

At one time nuclear energy was seen as the big hope for reducing the world's dependence upon fossil fuels, with their limited life expectancy (Source 1). Supporters of nuclear energy point out the big advantage it has over fossil fuels – no air pollution. Nuclear power stations do not emit greenhouse gases. They are not responsible for global warming. They generate electricity relatively cheaply. However, after the huge costs of building the power station are taken into account, electricity from nuclear power stations is usually more expensive than that from thermal power stations.

Source 2 shows the top nine producers of nuclear energy, plus India (see page 93) which currently lies seventeenth. Although most of the leading producers are MEDCs, this is set to change in the future. Many of the leading producers are not planning new nuclear power stations because of public fears over safety, and high building and decommissioning costs. Twenty-two of the last 31 nuclear power stations have been built in Asia. Of the 27 new ones under construction, 18 are in Asia.

Nuclear reactors produce radioactivity, which is dangerous to people and animal life. High levels of radioactivity cause cancers and death. Whilst safety standards are usually very high, there are occasional

| Source 1 | The life expectancy of fossil fuels in 2000 |

Fossil fuel	Number of years
Coal	230
Oil	41
Natural gas	62

leaks, which worry people living close by. Much more dangerous and frightening was the explosion at the Chernobyl power station in the Ukraine in 1986; its cloud of radioactive dust was carried for thousands of kilometres by the wind and reached as far west as the UK. Levels of radioactivity in waste from nuclear power stations take hundreds, or even thousands, of years to fall to levels that cannot damage people. As one member of Greenpeace said: 'The nuclear industry is unable to deal with the waste it has already created, let alone the waste it will create in the future.'

These reasons help to explain why it took so many years for planning permission to be given for Sizewell B, the last nuclear power station to be built in the UK. No new nuclear plants are planned. Instead, research and investment is directed towards alternative sources, such as wind and solar power.

| Source 2 | World nuclear power production and plants, 2004 |

Country	Electricity output from nuclear power (thousands of MW)	Number of reactors in use	Number of reactors under construction
1 USA	98.2	104	0
2 France	63.3	59	0
3 Japan	45.4	54	2
4 Russian Federation	20.7	30	3
5 Germany	20.6	18	0
6 Korea	15.8	19	1
7 Canada	12.1	17	0
8 United Kingdom	12	27	0
9 Ukraine	11.2	13	4
17 India	2.5	14	8

Nuclear energy
India

Although only seventeenth in the world (Source 2, page 92), as a producer of nuclear power India is one of a number of Asian countries who are leading the construction of new nuclear power stations worldwide. India and China are the most populous countries in the world, each with over a billion inhabitants. As demand for electricity increases, both are looking to move away from their dependence on coal-fired thermal power stations. Nuclear power is seen as the solution, and a much cleaner one emitting virtually no **greenhouse gases** – something which could help the **Kyoto Protocol** become effective in future years.

India's first nuclear power stations were the two Tarapur reactors opened in 1969. These were boiling water reactors (BWRs), the only ones of this type built in the country. They were followed by 12 pressurised heavy water reactors (PHWRs) opened between 1972 and 2000 (Source 1). India deliberately built nuclear reactors in areas furthest away from domestic coal supplies – two each in Kaiga, Madras, Kakrapar and Narora, and four at Rajasthan. These produced 2270 MW of electricity in 2004, 3.3 per cent of India's total.

India now has a further eight reactors under construction, all on existing nuclear sites (Source 1), with most of these being the new generation of fast-breeder reactors. By 2020, total production should be 20 000 MW. In the longer term, India plans to build a further 24. The industry today is regulated by the government via the Department of Atomic Energy (DAE) and run by the government-owned Nuclear Power Corporation of India Limited (NPCIL). Whilst its first two reactors were foreign-built, the DAE and NPCIL now carry out the commissioning, design, manufacture, construction and operation of nuclear power stations using Indian expertise and labour.

Source 1 India's nuclear power stations

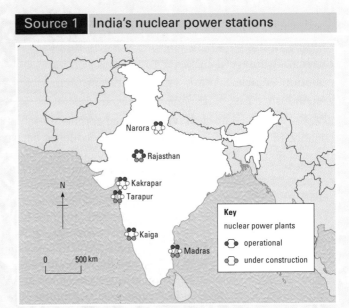

Source 2 Tarapur nuclear power station, India

1 a Make a copy of the table below and name three examples of jobs in each of the sectors named.

Primary	Secondary	Tertiary	Quaternary
1	1	1	1
2	2	2	2
3	3	3	3

b Write out and complete each of the following sentences to give definitions for the four sectors of economic activity.
  i Primary industries involve …
  ii Secondary industries involve …
  iii Tertiary industries involve …
  iv Quaternary industries involve …
c The employment structure for Brazil is given below.
  Primary 20% Secondary 25% Tertiary 55%
  i Draw a pie graph to show employment structure in Brazil.
  ii Describe the differences between the employment structure in Brazil and the employment structures in Bangladesh and the UK.
  iii Suggest reasons for the differences you have described in **ii**.

2 Look at Sources 1 and 2 below.

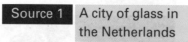

Source 1 A city of glass in the Netherlands

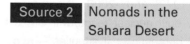

Source 2 Nomads in the Sahara Desert

a Describe the farming shown in each photograph.
b State the main differences between them.
c Suggest reasons why the farming is so different in the two areas shown.

3 a From Source 1 on page 92, state the evidence for each of the following.
  i India is less wealthy than the UK.
  ii India's population is increasing more quickly.
  iii People in India have a lower quality of life than people in the UK.
b Draw a labelled sketch map to show the location and different types of industry in India.
c An American multi-national company is investigating whether or not to open a factory in India for the first time. Write a report stating:
  • the advantages of locating a factory in India
  • the likely disadvantages
  • whether or not you would advise them to go ahead, giving your reasons.

4 | **Source 1** | A factory system

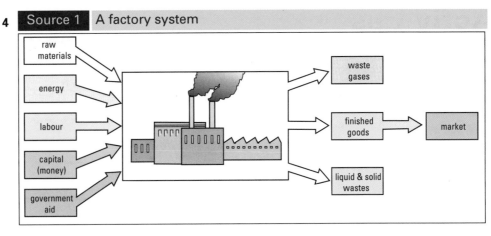

   a **i** Name two types of energy inputs used in factories.
       ii How many other different inputs are shown in Source 1?
       iii Name the output shown in Source 1 from which the factory makes its
         profits.
   b State the evidence from Source 1 that factories can cause pollution.
   c Describe how raw materials are changed into finished goods in factories,
      such as for making cars.

5 a | **Source 1** | Total world consumption of energy

Year	million tonnes of oil equivalent	Year	million tonnes of oil equivalent
1974	5 600	1989	7 800
1979	6 600	1994	8 000
1984	6 800	1999	8 500

    i Draw a graph to show the data in Source 1.
      ii How many more million tonnes of energy were consumed in 1999 than
        in 1974?
      iii State two reasons for the increase in world energy consumption
        between 1974 and 1999.
   b The percentages contributed by different energy sources in Europe in
      1999 were: oil 42% coal 22% natural gas 25% nuclear 8% HEP 3%.
     i Draw a divided bar graph to show these percentages.
     ii What percentage in Europe was from fossil fuels?
   c **i** 'Most non-renewable sources of energy cause major environmental problems'
     ii 'Most renewable sources of energy are clean and environmentally friendly.'
       Write a short report either supporting or disagreeing with each statement.

  6 a Plot the nuclear electricity output and number of reactors for each of the countries
      in Source 2 page 92 using pairs of bars on one graph.
    b **i** In 1995 the UK abandoned plans to build any more nuclear power stations. Give
       three reasons for this decision.
      ii India has embarked on a major programme to construct new nuclear power
        stations. Why have they made such a different decision from the UK?
    c Nuclear power is a controversial topic. It is supported by some governments, for
      example France, India and China, but fiercely opposed by organisations like
      Greenpeace and Friends of the Earth. Working in groups, write down arguments
      that both sides might use in a debate about nuclear power.

Development

Unit Contents

- Development indicators

- Different levels of development

- Development in a LEDC: Ethiopia

- Countries with different levels of development: Brazil and Italy

- Contrasts in development within countries: Italy and Brazil

- Growing economic prosperity: Malaysia, a newly industrialising country

- Declining economic prosperity in a MEDC: South Wales, UK

Rich and poor – where do you think the photographs were taken? The left one shows part of a ghetto in Harlem in New York City, and to the right is the impressive CBD skyline in São Paulo, Brazil. How typical of each country are these two scenes?

Development indicators

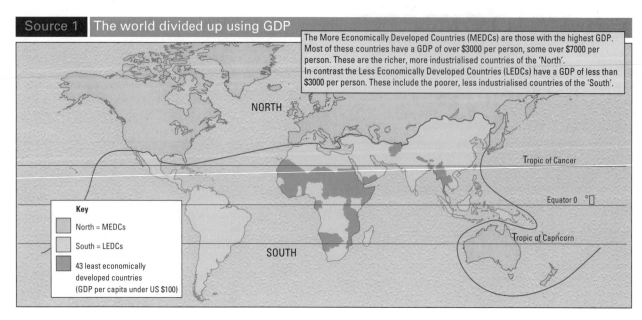

Source 1 The world divided up using GDP

The More Economically Developed Countries (MEDCs) are those with the highest GDP. Most of these countries have a GDP of over $3000 per person, some over $7000 per person. These are the richer, more industrialised countries of the 'North'.
In contrast the Less Economically Developed Countries (LEDCs) have a GDP of less than $3000 per person. These include the poorer, less industrialised countries of the 'South'.

NORTH

Tropic of Cancer

Equator 0 °

Tropic of Capricorn

SOUTH

Key

North = MEDCs

South = LEDCs

43 least economically developed countries (GDP per capita under US $100)

The meaning of development

What do we mean by development? In the past, it often meant how wealthy a country and its people were, quite easily measured by **Gross Domestic Product (GDP)**. Source 1 shows how the world can be divided into levels of development using this data. However, it is much more complex than this. Since 1990 when the first comprehensive set of **development indicators** were produced in the United Nations *Annual Human Development Report*, a much wider view of development has been adopted.

Economic factors are still important. If a country has a high GDP, it is likely that its people have a good quality of life, but such wealth is often unevenly distributed. It is also important how it is spent, for example building hospitals or buying weapons! However, human well-being depends on a wide range of factors like health, education and access to services like water, sanitation and electricity. Others like freedom, democracy, security and equal opportunities can be difficult to measure, yet are just as important.

Today, the Human Development Index (HDI), rather than GDP on its own, is used. It is based on:

- **Wealth (economic):** Gross Domestic Product (GDP) per capita measured in $US with an adjustment made for relative purchasing power (purchasing power parity or PPP) within each country
- **Knowledge (population):** an education index based on adult literacy and the total enrolled in primary, secondary and tertiary education
- **Health (social):** life expectancy at birth.

The final HDI figure is the average of the three indices within set minimum and maximum figures (Source 2). A world map showing global differences in HDI is on page 102.

Measuring development

A wide range of data is now used to help us measure levels of development. The United Nations annual report now includes over 30 tables containing more than 200 sets of data or **indicators** as they are called. Certain indicators are grouped together to cover a specific topic, for example poverty, trade and gender inequality. It is difficult to look at all these indicators in detail, so selected indicators are commonly used to assess economic, population and social development.

| Source 2 | Calculating the Human Development Index (HDI), from UNDP Human Development Report, 2003 |

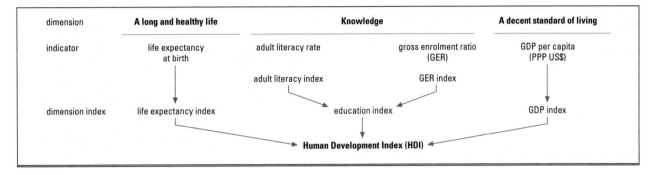

Economic indicators

As explained, GDP has traditionally been used to assess levels of human development and, although a wider range of indicators is now used, GDP is still important and is one of three measurements which are used to calculate overall HDI.

Source 3 lists some important economic indicators for ten selected countries at various levels of development (several of which are featured in more detail in this unit). The figures for GDP and energy consumption show huge differences between wealthy, high consuming More Economically Developed Countries (MEDCs) like Sweden and the USA compared

to Less Economically Developed Countries (LEDCs) like Kenya and Ethiopia.

Trade and industry are also important indicators. Source 3 shows imports and exports for each country as a percentage of the GDP. Generally, MEDCs have a balance between these figures, with exports sometimes contributing a higher proportion than imports. In LEDCs this is typically reversed, with imports accounting for a higher proportion. Employment structure – the proportion of the workforce employed in each sector (see page 59) – is also a good indicator of economic development. Wealthier countries usually have the highest proportion of workers in the tertiary sector.

| Source 3 | Economic indicators (from UNDP Human Development Report, 2003) |

Country	GDP $US ppp per capita	Annual energy consumption MWh per capita	Imports as % of GDP	Exports as % of GDP
Sweden	26 050	17 355	27	41
USA	35 750	13 241	14	10
United Kingdom	26 150	6 631	28	26
Italy	26 430	5 770	26	27
Malaysia	9 120	3 039	97	114
Brazil	7 770	2 122	14	16
China	4 580	1 139	26	29
Bangladesh	1 700	115	19	14
Kenya	1 020	140	30	27
Ethiopia	780	30	34	16

continued

Development indicators (contd)

Population indicators

Source 4 shows four of the most important population (**demographic**) indicators used to help measure development, for the same countries listed in Source 3 page 99. The figure for life expectancy is also used in the overall calculation for HDI (page 98). In MEDCs like the UK, most people can expect to live well into their late 70s, whilst poverty and poor health facilities are major factors in the significantly lower life expectancies for those at the bottom of the table.

Growth rate is calculated by subtracting the death rate from the birth rate. MEDCs usually have quite low birth rates – in Italy it is expected that this could become a negative figure in the next ten years. Less developed countries are beginning to see a fall in growth rates, but they are still several times higher than for developed countries. High rates of **infant mortality** are a major contributor to higher birth rates in LEDCs. As health services and general quality of life improve, both growth rate and infant mortality should fall, whilst life expectancy increases.

The fourth indicator in the table is the percentage of people living in urban areas.

Although it shows higher proportions for MEDCs, many LEDCs are seeing an increase in urban populations as people move to cities in search of work and what they believe to better opportunities. Most of the world's fastest growing cities are now found in LEDCs, particularly South America (see page 106).

Social indicators

Source 6 shows a range of social development indicators, the most important of which is the Human Development Index (HDI) introduced on page 99. This includes data taken from the last three columns on literacy and education. Countries with a high HDI score over 0.8, those with a medium HDI are between 0.5 and 0.79, whilst those under 0.5 have a low HDI.

Some countries score a lower HDI than might be expected when compared to their overall wealth measured by GDP. This is often because wealth is not shared evenly and people do not have access to services like education. Other social indicators in Source 6 show the that the populations of MEDCs like the USA have on average almost three times as much food per person as Ethiopia and a far better ratio of doctors per person.

Source 4	Population indicators (from UNDP Human Development Report, 2003)		
	Growth rate 1975–2002	**Infant mortality per 1000 births**	**Life expectancy in years**	**Urban population %**
Sweden	0.3	3	80	83
USA	1.0	7	77	80
United Kingdom	0.2	5	78.1	89
Italy	0.1	4	78.7	67
Malaysia	2.5	8	73	63
Brazil	1.8	30	68	82
China	1.2	31	70.9	38
Bangladesh	2.4	51	61.1	24
Kenya	3.1	78	45.2	38
Ethiopia	2.7	114	45.5	15

Source 5 Urban housing, Massachusetts

Source 6 Social indicators (from UNDP Human Development Report, 2003)

	HDI	HDI rank (177)	Doctors per 100 000 people	Average calories per day	Adult literacy %	% in primary education	% in secondary education
Sweden	0.946	2	287	3160	–	100	99
USA	0.939	8	279	3642	–	93	85
UK	0.936	12	164	3237	–	100	88
Italy	0.920	21	607	–	98.4	100	95
Malaysia	0.793	59	68	2899	88.7	95	69
Brazil	0.775	72	206	2938	86.4	97	72
China	0.745	94	164	2844	90.9	93	–
Bangladesh	0.509	138	16	–	41.1	87	44
Kenya	0.488	148	14	–	84.3	70	24
Ethiopia	0.359	170	1.4	1845	41.5	46	15

Different levels of development

Unit 4.1 looked in detail at a range of development indicators covering economic, population and social factors, with data for ten selected countries. The countries selected cover many different levels of development. Different indicators used together can give a fairly accurate picture of the level of development for individual countries (Source 1).

Source 2 maps HDI levels across the world and reveals quite a distinctive pattern of distribution. It roughly divides the world into two – the wealthier north where most MEDCs are found, and the poorer south where most LEDCs are located. In recent years the economies of a number LEDCs, such as Malaysia and Brazil, have grown rapidly, leading to the title 'Newly Industrialised Countries'. Whilst NICs are very much countries on the way up, the category LLEDC refers to the very poorest. The majority of these Less Less Economically Developed Countries, like Ethiopia (see pages 104–5), are in Sub-Saharan Africa.

Source 1	Development terms and levels	
MEDC	More Economically Developed Country	GDP over $10,000 per capita*, HDI over 0.8, high life expectancy
NIC	Newly Industrialised Country	GDP $3–10,000 per capita*, HDI approx 0.65–0.8, good life expectancy
LEDC	Less Economically Developed Country	GDP $1000–$3000 per capita*, HDI approx 0.45–0.65*, low life expectancy
LLEDC	Less Less Economically Developed Country	GDP less than $1000 per capita, HDI under 0.45, very low life expectancy

*Figures given are approximate/guidelines only

Source 2	Global Human Development Index (HDI)

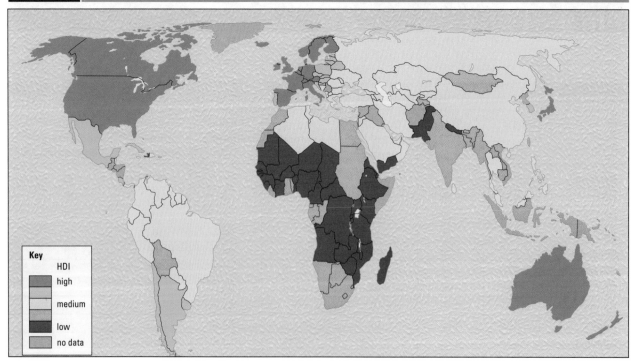

Key

HDI

high

medium

low

no data

Source 3 | Floods in Bangladesh

Why is development so unequal?

What are the reasons for such wide differences in development? Why is there such a big 'development gap'?

- Many LEDCs are located in the Tropics. This region experiences up to 100 tropical storms or hurricanes every year (pages 48–9), causing widespread damage. Earthquakes and volcanoes are also major hazards.
- Extreme climatic conditions or changes in normal weather patterns often case widespread flooding or drought, especially in monsoon areas on the edge of the Tropics or in the arid, desert regions just outside the Tropics (Source 3).
- Hot, wet conditions are excellent breeding grounds for disease-carrying insects like mosquitoes and for the spread of bacteria and viruses.
- **Colonisation** of many countries in Africa, South America and Asia by European powers since the sixteenth century has left a long-lasting legacy. Many colonisers used raw materials from the countries they claimed to build their own industries, to the detriment of

that country. They also established trade patterns, some of which still exist today.
- Many LEDCs have been encouraged to borrow large sums of money from MEDCs and major global organisations. Paying back even the interest of these debts can account for up to a third of the country's GDP.
- Lots of LEDCs are fairly newly independent. Many suffer from political instability and frequent conflict and civil war between different groups, preventing economic development.

Difference within countries

Not only do development levels vary between countries, they also vary within countries. The case study on pages 108–9 looks at differences in Italy and Brazil. In both countries there are considerable differences between regions. Although development indicators are very useful to allow comparisons of development levels between countries, it is very important to remember that these are 'average' figures. Country data does not show regional variety.

Development in a LEDC
Ethiopia

Ethiopia is the only country in Africa not to have been colonised. It was one of the centres of ancient civilisations, at the north end of the Great East African Rift Valley. Most of the country is mountainous, the source of the Blue Nile which flows north through Sudan into Egypt. Whilst the lowland areas in the south-east and north-east have frequently suffered from prolonged drought and famine, there are fertile valleys amongst the mountains where farming takes place.

Ethiopia has a population of over 66 million, the third highest in Africa. The country has been **landlocked** since 1993, when Eritrea became a separate independent country, two years after a long period of civil war. From 1998 to 2000, Ethiopia was at war again, this time with Eritrea. In 2002 the position of the border between the countries was finally agreed. However, the constant cycles of war and drought have had a major impact on the lives of Ethiopians. This is shown clearly in the various development indicators shown on the tables on pages 99–101.

Ethiopia was ranked 170 out of 177 countries in 2002, according to the Human Development Index (HDI), making it one of the poorest and least developed countries in the world. Part of the world's least developed region, the Sub-Saharan region of Africa

| Source 1 | Ethiopia and Sub-Saharan Africa |

(Source 1), it can be classed as a Less Less Economically Developed Country or LLEDC. These are countries with a GDP of under $1000 per capita and some of the lowest life expectancy figures in the world.

| Source 2 | Population indicators (from UNDP Human Development Report, 2003) |

Indicator (used to calculate HPI)	Ethiopia (% unless stated)	Sub-Saharan Africa (millions)
Living on less than $1 (PPP US$) per day	26.3	323
Total population undernourished	n.a.	185
Primary age children not in school	54	44
Primary age girls not in school	59	23
Children under age 5 dying each year	171 (per 1000 live births)	5
People without access to improved water	76	273
People without access to adequate sanitation	88	299

With a figure of 55 per cent, Ethiopia is also only ranked 92/95 developing countries on the Human Poverty Index (HPI) introduced in 1997 (Source 2). Only four other countries have an HPI over 50 per cent – Mali, Burkina Faso, Niger and Zimbabwe. Even within Sub-Saharan Africa, Ethiopia is very poor as the figures in Source 2 show.

During the most recent war with Eritrea, the proportion of GDP spent on defence and weapons was over 13 per cent – a rise of over 10 per cent from 1997. This is a higher percentage than that spent on education and health combined, and a severe drain on the economy. This has fallen following the peace agreement, but now the country is suffering from yet another prolonged period of drought. It also has the third highest number of people with HIV/AIDS in the world.

Despite the hardships and difficulties, Ethiopia is making progress, especially in terms of some social aspects of human development. Source 3 shows some of these, part of a government plan of reform. Some of this has been helped by decentralising to regions, but also through aid from the World Bank and International Monetary Fund.

Ethiopia also has a growing sugar industry employing over 10 000 in just one valley. There is potential for this to develop further, providing additional income and employment. However, it has to compete globally with wealthier countries who often support their own sugar industries with subsidies and tariffs.

Source 3	Improvements in human development	
	% 1990	**% 2002**
School enrolment	33	64
Youth literacy	42	56
Adult literacy	28	40
Infant mortality	128	116
GDP per capita		<5.6% since 1995

| Source 4 | Sugar cane farming in Ethiopia |

Countries with different levels of development
Brazil and Italy

Can you name any wealthy and poor areas within your local area? Differences in development can be seen within a single city, between regions in a country and between countries. Remember how the photographs on page 97 were the opposite of what we might have expected. Here we compare Brazil and Italy. A variety of indicators are used to measure the level of development:

- economic indicators, for example GNP, trade and aid
- social indicators, for example health care and education.

Source 1 | **Brazil**

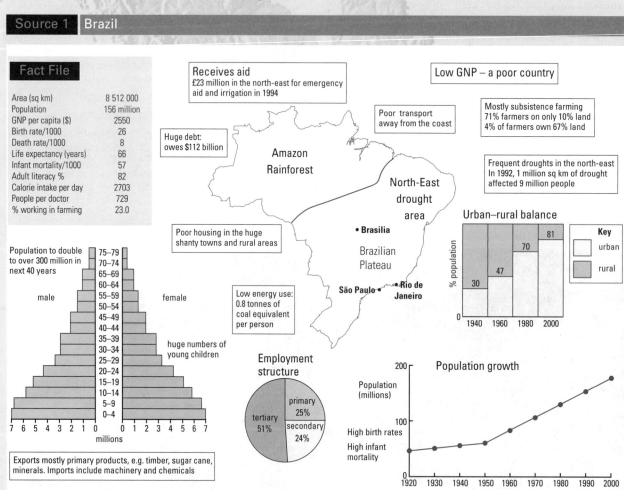

Fact File

Area (sq km)	8 512 000
Population	156 million
GNP per capita ($)	2550
Birth rate/1000	26
Death rate/1000	8
Life expectancy (years)	66
Infant mortality/1000	57
Adult literacy %	82
Calorie intake per day	2703
People per doctor	729
% working in farming	23.0

Receives aid
£23 million in the north-east for emergency aid and irrigation in 1994

Low GNP – a poor country

Poor transport away from the coast

Mostly subsistence farming
71% farmers on only 10% land
4% of farmers own 67% land

Huge debt:
owes $112 billion

Amazon Rainforest

Frequent droughts in the north-east
In 1992, 1 million sq km of drought affected 9 million people

North-East drought area

• Brasilia

Brazilian Plateau

São Paulo • • Rio de Janeiro

Poor housing in the huge shanty towns and rural areas

Urban–rural balance

Key
urban
rural

% population: 30 (1940), 47 (1960), 70 (1980), 81 (2000)

Population to double to over 300 million in next 40 years

male / female

huge numbers of young children

Low energy use: 0.8 tonnes of coal equivalent per person

Employment structure
primary 25%
secondary 24%
tertiary 51%

Population growth
Population (millions)
High birth rates
High infant mortality

Exports mostly primary products, e.g. timber, sugar cane, minerals. Imports include machinery and chemicals

Brazil: summary

Brazil is one of the world's less economically developed countries (LEDCs). The figures for birth rate, energy consumption and infant mortality support this. Brazil also has some characteristics expected of the MEDCs. These include the low death rate, the percentage of people living in urban areas and a trade surplus.

Since the 1960s, there has been an economic miracle in Brazil. Large-scale industries have developed, raw materials and energy resources have been exploited and the GNP has risen. However, much of the country remains poor and undeveloped, especially the north-east where there are frequent droughts. The government also borrowed heavily to pay for these developments.

Source 2 Italy

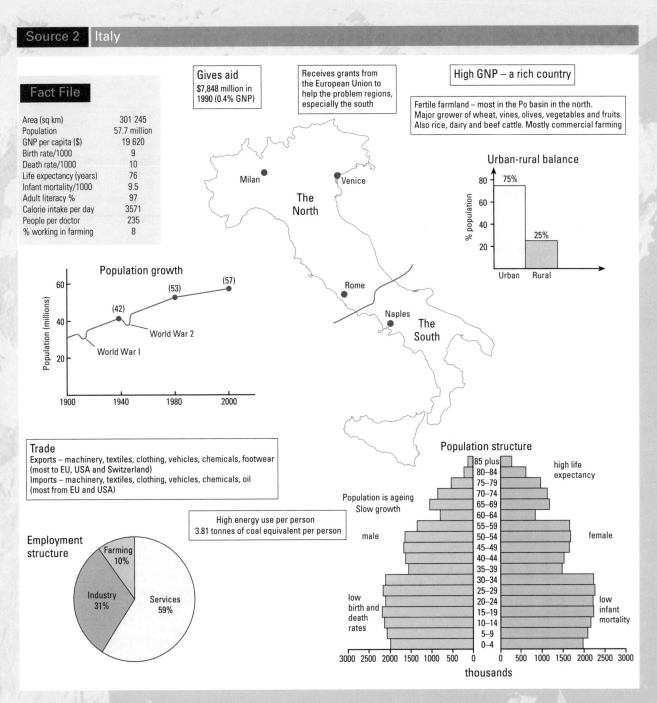

Fact File

Area (sq km)	301 245
Population	57.7 million
GNP per capita ($)	19 620
Birth rate/1000	9
Death rate/1000	10
Life expectancy (years)	76
Infant mortality/1000	9.5
Adult literacy %	97
Calorie intake per day	3571
People per doctor	235
% working in farming	8

Gives aid
$7,848 million in 1990 (0.4% GNP)

Receives grants from the European Union to help the problem regions, especially the south

High GNP – a rich country

Fertile farmland – most in the Po basin in the north. Major grower of wheat, vines, olives, vegetables and fruits. Also rice, dairy and beef cattle. Mostly commercial farming

Milan
Venice
The North
Rome
Naples
The South

Urban-rural balance
% population
75%
25%
Urban Rural

Population growth
Population (millions)
(42) (53) (57)
World War 2
World War I
1900 1940 1980 2000

Trade
Exports – machinery, textiles, clothing, vehicles, chemicals, footwear (most to EU, USA and Switzerland)
Imports – machinery, textiles, clothing, vehicles, chemicals, oil (most from EU and USA)

Employment structure
Farming 10%
Industry 31%
Services 59%

High energy use per person
3.81 tonnes of coal equivalent per person

Population structure
85 plus
80–84
75–79
70–74
65–69
60–64
55–59
50–54
45–49
40–44
35–39
30–34
25–29
20–24
15–19
10–14
5–9
0–4
high life expectancy
Population is ageing
Slow growth
male
female
low birth and death rates
low infant mortality
3000 2500 2000 1500 1000 500 0 500 1000 1500 2000 2500 3000
thousands

Italy: summary

Italy is one of the world's more economically developed countries (MEDCs). It has none of the characteristics of a LEDC except a small trade deficit. Italy has generally high living standards, low population growth and a high GNP. Poorer Italians are supported by welfare payments.

The country has a variety of landscapes and climates. There are large areas of fertile farmland which are used intensively to grow arable crops, vines and vegetables. The country can afford to import oil and has supplies of gas, oil and HEP (hydro-electric power). There is a long history of manufacturing and there are many different industries.

Italy also has problems. There are summer droughts and several volcanoes in the south, and there may be avalanches and the occasional earthquake in the Alps. The world recession has caused a trade deficit in recent years. Congestion and pollution are growing problems in large cities such as Milan, Venice and Rome. Perhaps Italy's greatest problems lie in the south of the country which remains quite undeveloped in many areas.

4.5 Contrasts in development within countries
Italy and Brazil

Wealth is not shared evenly within a single country. It is often concentrated in just one favoured region called the **core**, leaving other regions quite poor in comparison. These poorer regions are called the **periphery**.

Italy

Italy is a country with a north–south divide (Source 1). The north, especially the Po basin, is the core region and is wealthier and more developed than the south. The south of Italy, called the *Mezzogiorno,* is the periphery.

| Source 1 | Italy's north–south divide |

Advantages of the north:

- supplies of natural gas in the Po basin and HEP from the Alps
- more jobs in industry and services
- fertile lowland with irrigation water available
- large cities, for example Milan, Turin and Genoa connected by an efficient transport system
- close to large European markets
- better-quality housing and services and higher standard of living

Disadvantages of the south:

- mountainous relief makes communications difficult
- the climate is hot and dry in summer with a few months' drought
- heavy winter rainfall causes soil erosion and flooding
- the rocks are mostly limestone and form thin soils
- low yields of wheat, olives and vines
- poor-quality grazing for sheep and goats
- poor transport, little industry, emigration

Since 1950 the Italian government has invested money to try to improve the south. In recent years the EU has also provided grants and loans. In the south:

- some new *autostradi* (motorways) have been built
- new irrigation schemes allow tomatoes, citrus fruits and vegetables to be grown
- some large-scale industry, such as iron and steel, and car manufacture, has located in the South.

However, the north–south divide remains and the gap is widening.

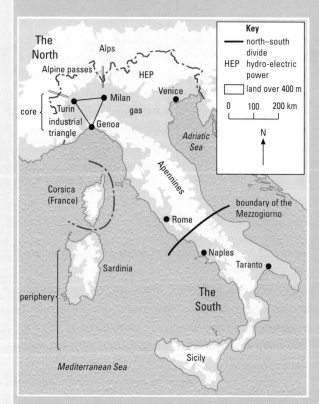

	North Italy	**South Italy**
Area %	60	40
Population %	63	37
Birth rate per 1000	11	17
Death rate per 1000	10	8
Income per person (million lira)	>2500	<1600
% farm production	65	35
% share of hospital beds	74	26
% unemployment	8	22

Brazil

Brazil's core region is the south-east of the country. The north and north-east form the periphery (Source 2).

| Source 2 | Brazil's south-east–north-east divide |

Prosperity and urban growth in the south-east of Brazil

The north-east

The north-east forms part of the periphery in Brazil. It is poorly developed and has many problems.

- The region suffers frequent droughts. In 1992 the drought affected 9 million people.
- Most farmers are subsistence farmers.
- The land is poor with infertile and eroded soils.
- Crop yields are low but the birth rate is high – there is not enough food to feed the population.
- The best land is used for plantations, often owned by **transnationals**. The crops are for export.
- The region has poor housing and services.
- Thousands of people have migrated from the area.

The south-east

Early growth was linked to coffee-growing near São Paulo. Coffee, gold and diamonds were exported. Later, rapid growth began with the mining of iron ore, the making of steel and the manufacture of cars and ships. Services were provided for the growing number of **migrants** from the rural areas.

Why is the south-east the core region?

- It is the centre of commerce, industry, education, transport and culture.
- The region has the highest standard of living in Brazil and contributes most to GNP.

However, many people live in shanty towns and there is major congestion and pollution in the cities.

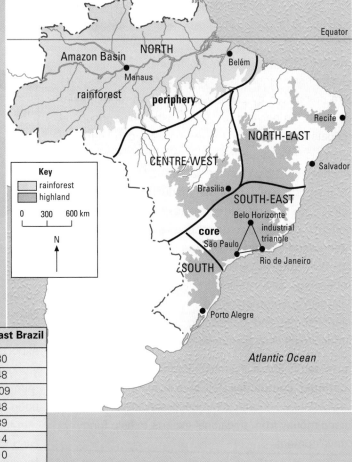

	South-East Brazil	North-East Brazil
Population %	42	30
Birth rate per 1000	22	48
Infant mortality per 1000	49	109
Life expectancy (years)	63	48
Adult literacy%	72	39
% share of national wealth	64	14
% employed in industry	70	10
% with clean water	64	23

4.6 Growing economic prosperity
Malaysia, a newly industrialising country

Many of the countries in South-East Asia named in Source 1 have fast-growing economies. Much of this growth is due to industrialisation. China and India have their own huge populations which provide a ready market. South Korea, Taiwan, Hong Kong and Singapore are known as the 'Tiger economies' because of the speed with which they have industrialised and developed over the last 50 years. Industry has grown, and continues to grow, in other countries of the Far East, such as Malaysia, which is why 'newly industrialising countries' is a good label for them.

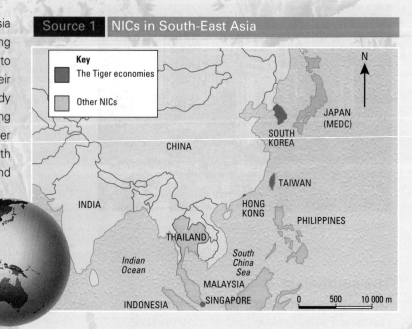

Source 1 NICs in South-East Asia

Key
■ The Tiger economies
□ Other NICs

Source 2 shows just how Malaysia's industry has changed in 30 years. In the 1970s exports were dominated by primary products like rubber, palm oil, cocoa, timber and tin. Today, although agricultural products are still important, exports are dominated by electronic goods plus chemicals, petrol and natural gas. Why have such big changes happened?

Many LEDCs rely on primary industry, especially the export of agricultural products, to support the economy. However, growers receive little income from unprocessed raw materials. It is the importing country and its workers who make most money from processing and selling finished products. The Malaysian government were determined to develop its manufacturing and processing industries. In 1970 it introduced its New Economic Policy (NEP). This:

- provided financial incentives for foreign **trans-national companies** (TNCs) to invest in Malaysia
- trained its workforce in the necessary skills
- used money from traditional exports to help fund development.

So successful was the NEP that in just 20 years only Japan and the USA were producing and exporting more

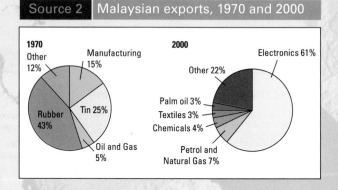

Source 2 Malaysian exports, 1970 and 2000

1970
Other 12%
Manufacturing 15%
Tin 25%
Rubber 43%
Oil and Gas 5%

2000
Electronics 61%
Other 22%
Palm oil 3%
Textiles 3%
Chemicals 4%
Petrol and Natural Gas 7%

electronic goods than Malaysia. However, some of this success was because workers were paid low wages, allowing them to produce goods far more cheaply than most MEDCs. The money earned by these new exports has not helped everyone – it is mainly those working on the mainland who have benefited, whilst many living on the islands are still dependent upon farming.

In the 1980s Malaysia also started to develop its own car industries, having worked since the 1960s with foreign companies, importing components and assembling them. Components were now made in Malaysia and in 1985 the Malaysian car company

Proton was launched as part of the National Car Project (NCP). By the 1990s it was producing over 100 000 vehicles a year and exporting across the world. Government backing and very high **tariffs** on imported cars both helped the industry grow and gave it considerable advantages over the competition.

Source 3 shows the proportions employed in each sector of industry. At 40 per cent, tertiary industry is now the most important sector. Malaysia is South-East Asia's major tourist destination with over 7 million visitors a year. It is also developing other tertiary industries in a zone stretching south from the famous Petronas Towers (Source 4) in the capital city, Kuala Lumpur. This is centred on the development of 'Cyberjaya', a new garden city covering 750 km2 providing a base for high-tech industry, services and finance, for both Malaysian companies and foreign TNCs. It is an important part of the government's aim to move Malaysia from its relatively new status of an NIC to that of an MEDC by 2020. If this is successful, it will have grown from an LEDC to an MEDC in just over 50 years.

| Source 3 | Malaysia employment structure, 2000 |

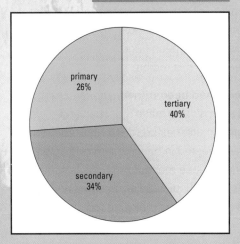

| Source 4 | Petronas Towers, Kuala Lumpur, one of the world's tallest buildings |

Declining economic prosperity in a MEDC
South Wales, UK

There was a time when the industrial landscape of Wales was littered with chimney stacks and smoke, signs that the region was dominated by heavy industry.

During the 1920s there were over a quarter of a million coal miners in South Wales. The number of miners and collieries declined dramatically, as Source 1 shows. Coal was a major source of fuel and helped provide the power needed for the Industrial Revolution. South Wales also had the raw materials needed to make steel: limestone, iron ore and coal (Source 1). Coal and steel were the two biggest industries in the region and part of their success was due to the fact that Britain still had an empire, which was a ready market for coal and steel.

Source 1 The decline in coal-mining in South Wales

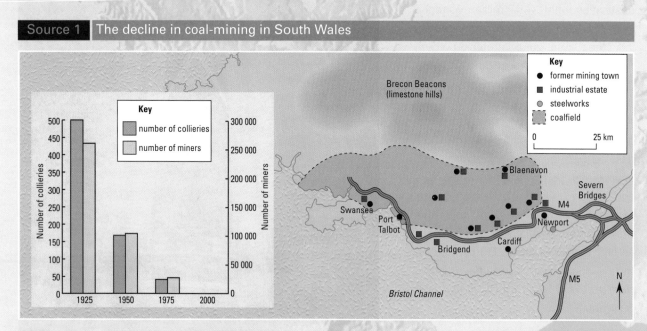

Source 2 Big Pit in Blaenavon

The graph (Source 1) shows just how rapid the decline of coal mining has been in South Wales, falling from 500 mines in 1925 to less than 50 in 1975 – with a loss of 200 000 jobs. Today there are just a few mines remaining, all privately owned. In many small villages in the region, mining was the main or only source of work, so the impact of mine closures has been huge.

Why did coal mining decline in this region?

- **The most accessible coal had been mined.** The remaining coal was in narrow, twisted seams. This made it difficult and expensive to mine as large machines could not be used. Government money was invested in other coal-mining areas. South Wales mines began to lose money.
- **The market for coal declined** especially as coal was no longer needed to power ships and trains. Its main use today is to make electricity.

- **Competition** came from countries like South Korea who can mine and sell coal more cheaply. Britain now imports most of its coal.
- **The development of alternative energy sources** – oil, gas and nuclear power - to produce electricity has led to a decrease in demand for coal.

Big Pit in Blaenavon (Source 2), once a working mine, has now been redeveloped as a tourist site, celebrating the history of coal mining in South Wales.

As the coal industry declined, so did iron and steel production. Today the only steelworks is found on the coast at Port Talbot. The raw materials needed (iron ore and coal) are now imported. The works is one of the most modern in Europe, using high-tech machines, employing very few people when in the past the industry employed thousands. Exhaustion of raw materials and overseas competition has hastened the decline. Steel is produced more cheaply in other countries and these now dominate the world market.

Regeneration of industry in South Wales

The decline of two major industries saw high unemployment levels in the region throughout the 1970s and 1980s. However, new industry has gradually taken the place of the old, heavy industries. Many of these are found on **industrial estates** or **science/business parks**. Imperial Science Park, part of the larger Celtic Lakes Business Park, opened near Newport in the mid 1990s (see pages 78–9).

Parts of South Wales were designated as **Development Areas** and **Enterprise Zones**, with funding been from the EU and/or UK Government. This meant that money was available to help establish new industry, channelled through the **Welsh Development Agency**.

Enterprise zone funding saw the building and development of five 'parks' on the coast at Swansea — the largest creating over 3000 new jobs with 300 small companies. Other areas were developed as leisure, shopping, heritage, tourism and residential areas. Amongst incentives to locate here was exemption from local taxes, assistance with planning applications and the availability of grants and low rate loans.

Many new developments are mainly located on the southern edge of the coalfield, where there is:

- easy access (good roads, connections to the rest of Britain and Europe)
- pleasant surroundings (greenfield sites)
- services / infrastructure provided
- cheap land with room to expand
- financial incentives
- nearby skilled workforce.

| Source 3 | Mid-Glamorgan Science Park, South Wales |

| Source 4 | The M4 at Port Talbot |

1 **a** What does GDP stand for? What does it measure?
   b Use the GDP data in Source 1 on page 99 for this exercise.
      List those countries with a GDP per capita (**i**) above $20 000 and (**ii**) below $10 000.
   c Using the two lists from (**b**), say which list is made up of MEDCs and which is made up of LEDCs.
   d Give two reasons why GDP is not always a good indicator of development.
   e Why is HDI a better measure of development than GDP?

2 Scattergraphs can show whether there is a link between GDP and the other indicators of development. Draw scattergraphs between GDP and some of the other indicators shown in Sources 1, 2 and 4 on pages 98–101. Source 1 below shows the example of GDP and birth rate. Write a sentence below each of your graphs to say what it shows.

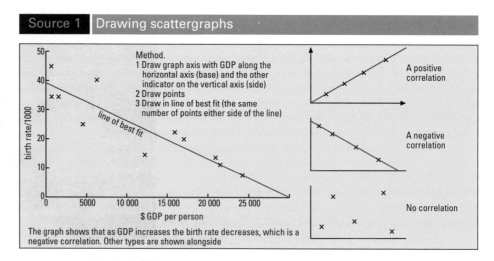

Source 1   Drawing scattergraphs

Method.
1 Draw graph axis with GDP along the horizontal axis (base) and the other indicator on the vertical axis (side)
2 Draw points
3 Draw in line of best fit (the same number of points either side of the line)

A positive correlation

A negative correlation

No correlation

The graph shows that as GDP increases the birth rate decreases, which is a negative correlation. Other types are shown alongside

3 **a** Produce a table in summary form to show the differences between Brazil and Italy. Use the following headings as a guide: Population characteristics; Employment (farming, industry and services); Trade; Urban rural balance; Energy use; Food intake; Aid; Problems; GDP; Literacy.
   b In what ways is Brazil as developed as Italy?
   c In what ways is Italy like a less economically developed country?

4 Italy and Brazil both have differences in development within their countries. You could use these questions for one or both countries.
   a What do you understand by the core and periphery in a country?
   b Draw a sketch map and label some of the main features of the country.
   c On your sketch map clearly mark the core and periphery.
   d Quote three statistics that show the differences between the two regions.

5 Study Source 2 which shows information about the UK compared with two countries in Central America.

   a Which indicator best shows the level of education?

   b Give two indicators that show the level of health care.

   c What does the term gross national product (GDP) per capita mean?

   d What is the relationship between GDP per capita and life expectancy in Source 2?

   e What are the advantages and disadvantages of using GDP per capita as a development indicator?

Source 2	Honduras, Nicaragua and the UK		
Indicators	**Honduras**	**Nicaragua**	**UK**
Population	5.8 million	4.1 million	58.9 million
GDP per capita	$600	$380	$18.342
Life expectancy	69 years	68 years	76 years
Access to safe water	87%	58%	100%
One doctor for every	1266 people	2000 people	850 people
Infant mortality	47 per 1000	51 per 1000	6 per 1000
Literacy	73%	66%	99%

6 a i Explain what is meant by the development gap.

     ii Give four reasons for this gap.

   b i Write short definitions for: MEDC, LEDC, LLEDC.

     ii Give an example of each.

7 a Why is Ethiopia classed as an LLEDC?

   b What are the main reasons for the poverty and low development levels within Ethiopia?

   c Using Source 3 on page 105, draw two pie charts comparing improvements in human development from 1990 to 2002.

   d With the two graphs from (c) to help you, describe the progress Ethiopia has made in recent years.

8 a Malaysia is an NIC. What does this mean?

   b What are the main differences between Malaysia's main exports in 1970 and those in 2000?

   c What actions did the government take in 1970 to help stimulate change?

   d What are Malaysia's plans for future economic development?

9 South Wales is an area of the UK which has seen its main industries decline.

   a What raw materials formed the basis for industrial development in South Wales?

   b i Look at the graph in Source 1, page 112. Describe how the number of miners and collieries has fallen.

     ii What are the main reasons for industrial decline?

   c What is happening in the region today to promote economic growth?

Migration

Unit Contents

Migration means the movement of people.
What makes people move?

Population change

Population change may mean an increase or a decrease in the number of people living in an area. There are three factors affecting population change: birth rate, death rate and migration. It is natural to assume that it means population increase because the world's population is continuing to grow at 1.3. per cent per annum (Source 1).

Birth rates

High **birth rates** contribute to population growth . Birth rate is the number of live births per 1000 people per year in a country or region. Source 2 shows the average birth rates for the different continents. Notice how much higher the average birth rate is in Africa, where there are many less economically developed countries (LEDCs), compared with Europe, where most of the countries are more economically developed (MEDCs). Some of the factors which help to explain these differences are given in Source 3.

Generally birth rates in the world are declining as more and more people are practising birth control. Wealth and education are the best contraceptives: richer and better educated people have fewer children. Why, therefore, is world population continuing to increase? It is necessary to look at the other element in the formula – the **death rate**.

Death rates

Low death rates also contribute to population growth. Death rate is the number of deaths per 1000 people per year in a country or region. Death rates are at an all time low (Source 2). Notice how much lower the death rates are than the birth rates. The reason for this is improved medical treatment and **primary health care,** which reduces the chances of a person becoming ill in the first place. Source 4 illustrates some of the ways in which this can be done. **Secondary health care** is more widely available. Particularly in the cities in LEDCs, there are hospitals to treat sick people, where modern medicines and drugs are available.

| Source 1 | World population growth 1950–2075 |

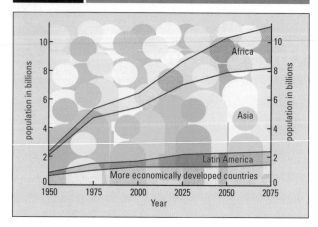

| Source 2 | Average birth and death rates |

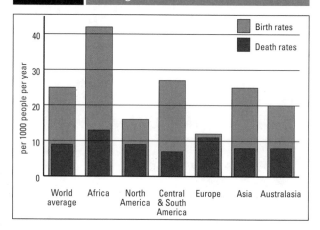

| Source 3 | Factors which affect the birth rate |

Factor	High birth rates in LEDCs	Low birth rates in MEDCs
Economic	• Children can work on the farm or earn money begging or selling goods in the city. • Children support elderly parents.	• Children cost their parents a lot of money. • Pensions for the old.
Social	• Little use of birth control. • 6–10 children in a family is normal.	• Many methods of birth control are used. • 2–3 children are the norm.
Political	Governments in Muslim and Catholic countries will not always provide family planning education.	Government-financed family planning services.

| Source 4 | Ways to prevent illness and death |

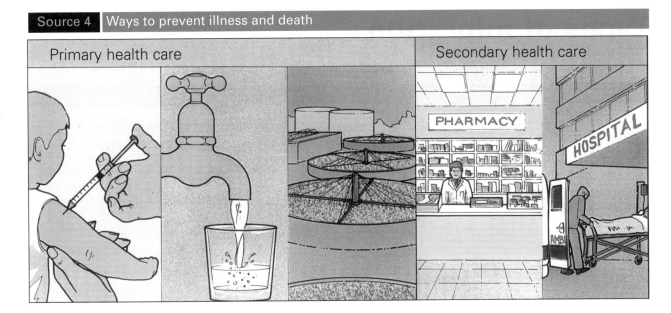

Primary health care | Secondary health care

Natural increase

The rate of **natural increase** in a country or region is worked out by the following formula.

$$\frac{\text{birth rate}}{\text{death rate}} = \text{rate of natural increase}$$

Using the values from Source 2 the rates of natural increase for the continents can be worked out.

The changing relationships between birth and death rates through time are shown on a graph called the demographic transition model (Source 5). What has really happened is that the death rate has fallen rapidly in the LEDCs in stages 2 and 3. It has fallen much more quickly than the birth rate. The big difference between the death rate and the birth rate has resulted in the high rate of population increase. Only in the MEDCs have birth rates fallen to the same level as death rates, so keeping the population increase low. Most European countries are in stage 4.

Migration

Together with birth and death rates, the population of a region or country is also affected by in-migration and out-migration. The difference between those moving in (immigrants) and those moving out (emigrants) is called the **migration balance**. International migration is measured using the **net migration rate (NMR)**. This is calculated giving the number of migrants per 1000 of the population. Figures from the late 1990s showed 60 countries where immigration was higher than emigration and 90 with higher emigration than immigration.

| Source 5 | The demographic transition model |

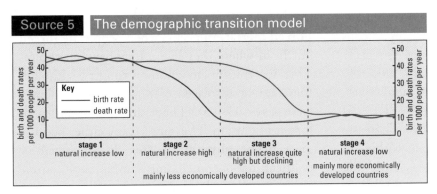

| Source 6 | World population increase, 1830–1999 |

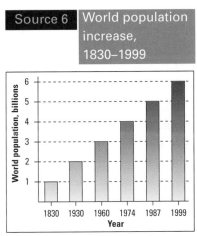

Types of migration

Source 1 | Types of migration

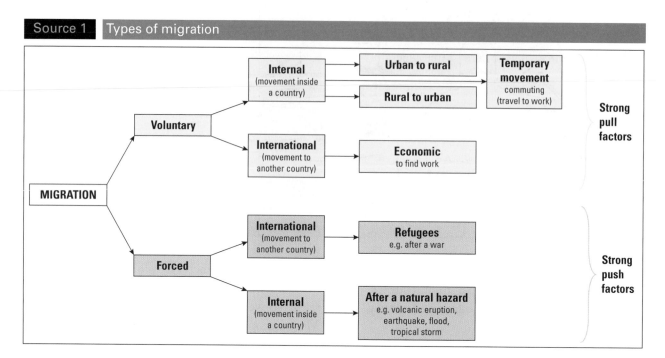

Migration is the movement of people, and throughout history people have moved from one place to another, inside their own country or to other countries. As Source 1 shows, there are many different types of migration, but they can be divided into two main groups – **voluntary** (by choice) or **forced** (compulsory). In either case this can be **internal**, within a country, or **international**, between countries.

Voluntary migration is where people choose to move inside their own country or emigrate to another country. The normal reasons for this are economic, to find work and increase their income and their quality of life. This usually means moving from rural areas to urban ones, especially in LEDCs. However, many people also move the other way, from urban areas to rural ones, although this is more common in MEDCs as people move away from crowded cities. There are also a number of people who move within an urban area (urban to urban), from near the suburbs to the city centre. **Commuters** do this on a daily basis from home to work.

There are many different kinds of **economic migrants**. The majority tend to be young adult males who move to find work. Sometimes migration or movement is **temporary** or **seasonal**, for example picking and harvesting crops. Seasonal migrant farm workers are often found within the European Union, or in US states like California (see pages 128–9). In Europe they were often called **guestworkers** who could earn far higher wages in another country, and sent much of what they earned home to their families. Temporary migrants are those who move from home for less than a year.

Forced migration occurs when people are **forced** out from where they live – they have no choice. This is typically to another country, although they may be displaced within their own country (see pages 124–5). There are many causes of forced migration. They are usually either because of a major physical disaster or for social or political reasons including warfare and ethnic cleansing. Forced migrants are called **refugees** (see pages 124–5).

Natural hazards like earthquakes, volcanic eruptions, hurricanes, landslides, flooding and drought are all physical reasons for having to move. In most cases, the victims and survivors of such disasters will move back when it is safe to do so or when homes and jobs are available again.

However, the biggest causes of **forced migration** result from the actions of people, especially war and persecution. Historically this includes large-scale migration such as Jews fleeing from Russia and Germany, and Palestinian Arabs ejected as Israel claimed more land. In recent years, wars in the Balkans, Afghanistan, Iraq, Rwanda, Sierra Leone, Somalia and the Sudan have added to the list of refugees, migrants and displaced persons.

Many recent wars have been civil warfare – factions within a country fighting one another. In some cases this has been in an effort to force out entire ethnic groups or communities – a process called **ethnic cleansing**. This was the case when the former Yugoslavia erupted into war and in

| Source 2 | A Peruvian village abandoned after landslides and flooding |

Rwanda, where the Hutus attempted to remove the Tutsis, leading to 800 000 deaths and 2 million refugees.

| Source 5 | People forced out from Gisenyi, Rwanda because of war |

Why people migrate

Unit 5.2 introduced some of the reasons why people migrate, especially those who are forced to do so. There is a wide variety of reasons why people either choose, or are forced, to move. These are known as **push–pull** factors, a theory introduced by Everett Lee in the 1960s. It identified a number of factors which were likely to 'push' people way from an area to migrate to another one. At the same time he also listed a number of 'pull' factors – reasons why migrants were attracted into a new area.

For some migrants, push factors are the most important, for others it is the pull factor. Sometimes it is a combination of the two. In rural areas there are many push factors at work. Especially in poorer LEDCs, rural areas can suffer from extreme poverty. There is often little work except for subsistence farming where drought or flooding can cause devastation. Such areas are often remote and isolated with few facilities such as schools or clinics. Basic services such as electricity, clean water and sanitation are often lacking.

Source 1   Reasons for rural to urban migration in LEDCs

Source 2   Reasons for urban to rural migration in MEDCs

Pull factors can be very powerful, especially for those living in rural areas in LEDCs, but also for those in MEDCs. The cities seem to offer well-paid jobs, better housing and services and a range of facilities including schools, hospitals, shops and entertainment. However, many of these pull factors are often just perceptions of what a place is like – the reality may be very different for a migrant arriving in a large city with no job and nowhere to stay.

Although people do still migrate from rural areas to urban ones in MEDCs, the recent trend is for urban to rural movement. Again there area a variety of push and pull factors. Push factors include wanting to move away because of noise, pollution, traffic congestion and high property prices. Pull factors, especially in MEDCs, include more space, attractive countryside, cheaper property prices (although not always) and a generally quieter lifestyle. Unlike many rural areas in LEDCs, those in MEDCs usually have all the basic services, although public transport and schools will be more restricted than in urban areas.

In MEDCs there has been an increase in urban–rural migration. Many people retire early and can move away from where they used to work, often urban areas. Others may work from home or are prepared to commute long distances to work in return for a better quality of home life in the countryside. This movement is called **counter-urbanisation** (see Units 6.8 and 6.9, pages 150–3).

Despite differences in migration in MEDCs and LEDCs, both types of population movement and change cause problems. Those moving from rural areas are often young adults, leaving behind an increasingly ageing population. Even if the area had good amenities, these may start to close as depopulation takes place, for example schools, shops, etc. Newcomers to a rural area may force house prices up beyond the reach of locals, causing conflict. Moving to an urban area may put increased pressure on space and resources, resulting in unemployment, increased crime and poor housing conditions (see Unit 6.4 and 6.7, pages 142–3 and 148–9)

Source 3 Many villages in MEDCs are losing local services like post offices

| Source 1 | Persons of concern to UNHCR, 1999–2003 |

End of year	Refugees	Asylum-seekers	Returned refugees	Others of concern			Total population of concern
				IDPs	Returned IDPs	Various	
1999	11 625 700	773 700	1 599 100	3 968 600	1 048 400	1 487 600	20 503 100
2000	12 062 100	949 300	767 500	5 998 500	369 100	1 653 900	21 800 400
2001	12 029 900	940 200	462 400	5 048 000	241 000	1 039 500	19 761 000
2002	10 594 000	963 500	2 426 000	4 646 600	1 179 100	953 400	20 762 600
2003	9 671 800	995 100	1 094 700	4 186 800	232 800	912 200	17 093 400

At the beginning of 2004, the United Nations High Commission for Refugees (UNHCR) estimated that there were 17.1 million 'persons of concern' to them across the world. Source 1 shows the breakdown of these groups over a five-year period. Source 2 shows the percentages for 2003.

Refugees are people outside their home country who cannot return for fear of persecution because of their race, religion, nationality, political opinion or membership of a particular social group. One hundred and forty-five countries accepted and signed the 1951 Geneva Refugee Convention and 1967 Protocol supporting this definition. Refugees have legal protection, the right to stay in the new country for as long as necessary and should be offered assistance with basic needs.

Asylum-seekers are people who have left their home country claiming persecution and looking for a place of safety where they can claim asylum and be granted refugee status.

| Source 2 | Persons of concern, percentages, 2003 |

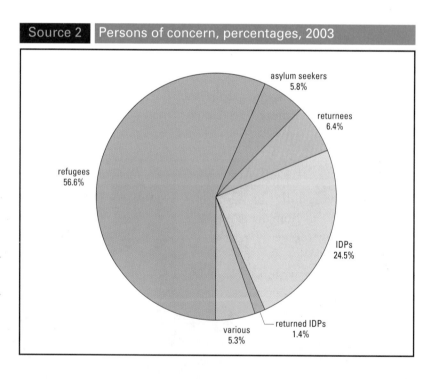

refugees 56.6%

asylum seekers 5.8%

returnees 6.4%

IDPs 24.5%

returned IDPs 1.4%

various 5.3%

The majority of refugees want to return home when it is safe for them to do so. This may be when a war has ended, although they may want to wait until basic facilities are available. This group of people are called **returnees**. To help they are often given transport home and grants for items such as tools and seeds. In very difficult circumstances, groups of returnees may be monitored.

Internally displaced persons (IDPs), like refugees, have been forced to leave their homes through persecution, often civil war, but have not left the country. They are persons causing concern because they do not have the rights of refugees but are often in very difficult situations. 'Various' includes the world's **stateless** people – approx 9 million globally.

Despite the high numbers still causing concern, Source 1 does at least show an encouraging trend with a 17 per cent decrease in numbers overall. The figure of 17.1 million is the lowest for ten years, with the total number of refugees falling from 10.6 million the previous year to 9.7 million in 2003. Source 3 shows the percentages of this 17.1 million by region. Overall, over a third of the people of concern to UNHCR are in Asia.

In 2003, 265 000 new refugees came from Sudan, Liberia, the Democratic Republic of Congo, Cote D'Ivoire and Somalia, some of the most recent refugee 'hot spots'. Many of these have fled civil war and major conflict within their own countries – an increasing proportion of refugees are from war zones. However, the aftermath of previous conflicts is still clear, with Afghanistan's 2.1 million refugees by far the biggest of any individual country of origin. This is 22 per cent of the total number of refugees worldwide. Pakistan, with an estimated 1.1 million, is the country hosting the highest number of refugees. Iran and Germany are home to approx 950 000 each, with 650 000 in Tanzania. Pakistan and Iran were the main country's Afghan refugees fled to, whilst drought, famine and war in several counties of East Africa has resulted in Tanzania's high total. LEDCs host the majority of the world's refugees, yet are the least able to provide the level of resources and care often needed to cope with mass exoduses of people who flee from home, often with little more than the clothes they are wearing.

| Source 3 | Persons of concern by region, percentages, 2003 |

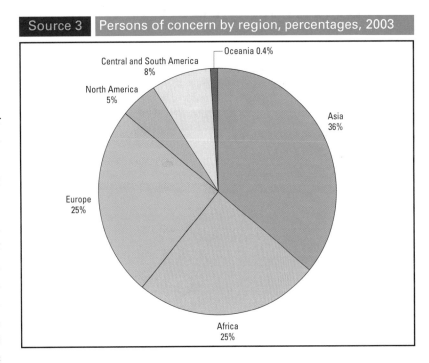

| Source 4 | In 2003, conflict in Sudan created a fresh exodus of refugees |

Internal migration in a MEDC
the UK

Source 1 shows the average population change for the different regions in Great Britain revealed by the 1991 census. Scotland and all the regions in the north of England lost people, while most regions in the south and east of England increased in population. This suggests that the long-established migration of people from the North to the South was still taking place (Source 2).

Source 1 | Population change 1981– 91

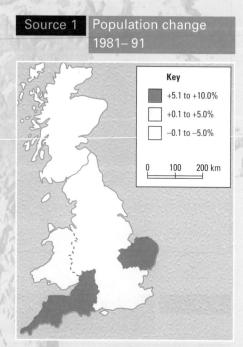

Key
- +5.1 to +10.0%
- +0.1 to +5.0%
- –0.1 to –5.0%

0   100   200 km

Source 2 | Migration movements

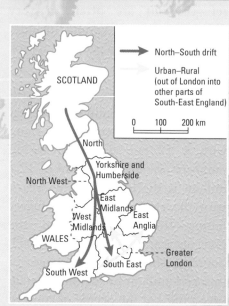

→ North–South drift

→ Urban–Rural (out of London into other parts of South-East England)

0   100   200 km

SCOTLAND
North
North West
Yorkshire and Humberside
East Midlands
West Midlands
East Anglia
WALES
Greater London
South East
South West

Source 3 | Population change by county 1981–91

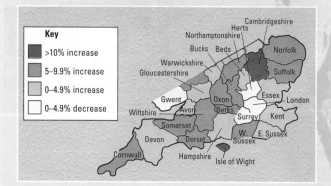

Key
- >10% increase
- 5–9.9% increase
- 0–4.9% increase
- 0–4.9% decrease

Cambridgeshire
Herts
Northamptonshire
Bucks   Beds
Warwickshire
Gloucestershire
Norfolk
Suffolk
Gwent
Oxon
Essex   London
Wiltshire   Avon   Berks
Surrey   Kent
Somerset
W. E. Sussex Sussex
Devon   Dorset
Cornwall   Hampshire
Isle of Wight

Source 4 | Job changes 1981–96

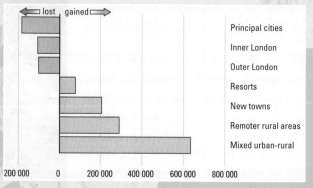

lost   gained

Principal cities
Inner London
Outer London
Resorts
New towns
Remoter rural areas
Mixed urban-rural

200 000   0   200 000   400 000   600 000   800 000

Source 3 gives more detail about population changes for southern England because it is based upon county statistics. It highlights more clearly the way in which Greater London has lost population, while counties to the north and west of London have grown the most. This change in population can be explained by the process of **urban to rural migration**. The push and pull factors for this are given on pages 122–3 and 152.

Many jobs have moved to rural areas (Source 4). Planners have also played their part in encouraging this movement. There are restrictions upon building new houses in the Green Belt which surrounds London. New towns, located outside London, were selected as growth points. Much of the population increase in Buckinghamshire is concentrated in the 'new town' of Milton Keynes.

Internal migration in a LEDC `5.6`
Brazil

Although patterns of movement in a country are always complex, two main migrations are dominant in Brazil (Source 1).

The main migration is from north to south. People living in the largely rural North-East and North of Brazil are much poorer than those in the South-East and South. The index of human poverty is highest in the North-East at 46 per cent and lowest in the South-East at only 14 per cent (Source 2). Poverty in the North-East has both human and physical causes. Most of the people are landless peasants. The large landowners pay low wages, offer work only at certain times of the year, and drive peasants off any empty land that they try to cultivate. Fewer workers are needed as a result of mechanisation on the farms. At the same time high birth rates are increasing population pressure on the land and the need for essential services to be provided is greater than ever. From time to time the interior of the North-East suffers from catastrophic droughts. When the crops fail, people are driven to the cities and farms in the South-East and South of Brazil in even greater numbers. Job-hungry migrants from the North-East are found everywhere in Brazil where there is a chance of work. Look at the case studies of São Paulo in Unit 6.2 pages 138–9 and Unit 6.4 pages 142–3 for more details about rural–urban migration.

The second movement is westwards, from the well-populated coastal region into the interior with its great empty spaces. Opening up the interior and tapping the wealth of mineral and timber resources of the Centre-West and North has been the dream of the Brazilian government for more than 50 years. The 'great march' westwards began with the creation of Brasilia as the capital in 1960. This was followed by a road building programme which is still continuing. Mining and logging companies moved in first, followed by cattle ranchers. Landless people, many of them from the North-East, followed in the hope of a fresh start. Amazon forest is still being cleared and new mineral finds are still being made. Several million Brazilians already live in the interior and more are moving there.

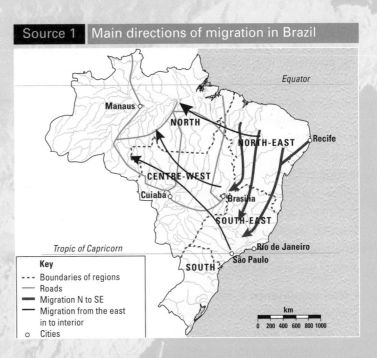

Source 1 Main directions of migration in Brazil

Key
- - - Boundaries of regions
— Roads
▬ Migration N to SE
— Migration from the east in to interior
○ Cities

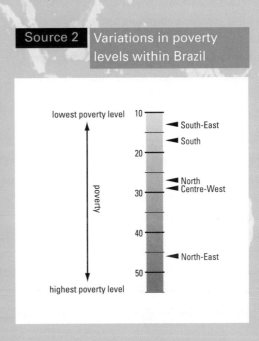

Source 2 Variations in poverty levels within Brazil

International migration
Mexico–USA

Given the differences in wealth and a shared border of 3300 km (Source 1), it is not surprising that there has been a long history of migration from Mexico north to the USA. The USA is a rich MEDC – it has a GDP of $36 000 and is placed 8/177 in the Human Development Index. In contrast, Mexico has a GDP of $8500 and an HDI ranking of 53/177.

The USA has actively encouraged migration from Mexico in the past, especially seasonal workers to pick crops or work in food processing factories in California. They classify migrants in three categories:

- **Non-immigrants:** foreign migrants entering the USA on a temporary basis – as students, seasonal workers or tourists. Visas are required for such visitors, issued for various time periods depending on their purpose
- **Lawful permanent residents (LPRs):** foreign-born individuals granted permission to live permanently in the USA via family or employment programmes or as refugees. Source 2 shows overall figures from 1986-2002, and the number and proportion legally accepted from Mexico
- **Undocumented migrants:** illegal or unauthorised immigrants entering without going through border checkpoints, using forged documents or out-staying their visa dates.

It is estimated that there are between 8 and 12 million illegal immigrants living in the USA. This figure is thought to be increasing at the rate of about half a million a year, with 50 per cent from Mexico. The length and often remote nature of the border makes it difficult to police, even though there are over 5000 Border Patrol Agents (Source 3) using trucks and helicopters around the clock. There are over 300 million legal two-way crossings each year. An estimated 2 million illegal crossings take place successfully, often at night.

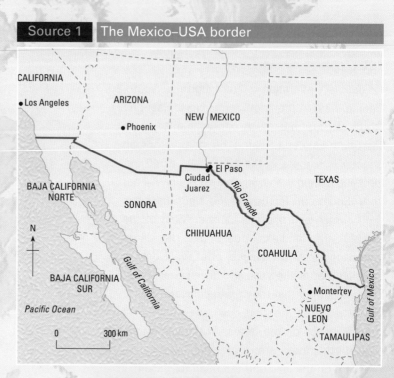

Source 1 The Mexico–USA border

Source 2 Legal immigration into the USA

Year	Total	From Mexico	
1986	601 000	66 000	(11%)
1987	601 000	72 000	(12%)
1988	643 000	95 000	(14%)
1989	1 090 000	405 000	(37%)
1990	1 536 000	679 000	(44%)
1991	1 827 000	946 000	(51%)
1992	973 000	213 000	(22%)
1993	904 000	126 000	(14%)
1994	804 000	111 000	(13%)
1995	720 000	89 000	(12%)
1996	915 000	163 000	(17%)
1997	798 000	146 000	(18%)
1998	654 000	131 000	(20%)
1999	646 000	147 000	(22%)
2000	849 000	173 000	(20%)
2001	1 064	206	(19%)
2002	1 063	219	(20%)

There are several push and pull factors encouraging Mexicans to migrate (see pages 122–3). Those living in poor rural areas are gradually seeing marginal farmland become more arid and worthless, making it increasingly difficult to make a living. They watch others leave, either to move to Mexico City or to find work and prosperity in the USA – further impacting on their own community. If they cannot get a visa to move legally, many feel they have little to lose by crossing illegally. If caught, they are normally returned immediately – and try again. Even without work visas, most can find work in the informal sector, especially in large cities like Los Angeles. Although low paid by US standards, most still earn far more than they would have at home. Money is often sent back to families in Mexico. However, finding somewhere to live is often more difficult, and many Mexican migrants (legal or not) live together in the poorest city neighbourhoods.

Ongoing talks between the presidents of Mexico and the USA are concentrating on reducing illegal immigrants partly by looking at a new temporary worker programme, with the chance of earning or qualifying later for LPR status.

Source 3	A Border Patrol Agent on the US side

Maquiladoras

Set up in 1965 by US companies eager to cut labour costs, these are factories mainly in the north of Mexico close to border where Mexican labour is used to assemble imported US components into finished goods before being exported for sale. Wages are much lower than in the US, but good for Mexico and skilled workers may get the opportunity to work in the US. Financial incentives such as subsidies and tax exemptions also encourage companies to set up, and there are now about 4 500 (not all US-owned) in Mexico producing electrical goods, clothing, auto parts and machinery.

Source 4	Mexicans working in a maquiladora

5.8 International migration
Vietnamese 'boat people'

Although the last of the Hong Kong camps housing Vietnamese boat people finally closed in June 2000, the flight, resettlement and/or repatriation of this particular group of international migrants remains an important one to study, covering a wide range of issues relevant to refugees and economic migrants.

Vietnam has had a turbulent history, occupied in turn by China, France and Japan. In 1954 the country split into two – North and South – ready for separate national elections in 1956. Instead, the leader of the south declared himself to be President and war broke out. In the early 1960s, during the Cold War between the world's leading superpowers – the USSR and USA – the USA started to move troops in to the south worried that the Communist north would overwhelm and claim South Vietnam. Years of bombing and fighting finally came to an end in April 1975 when US troops withdrew and the south finally fell. By the following year the new Socialist Republic of Vietnam was recognised.

Hundreds of thousands fled from the former South Vietnam, mostly by sea – the largest seaborne exodus of refugees in history. They took to almost anything that could sail. Hundreds crowded onto old, wooden boats (Source 2) in attempts to sail to neighbouring countries. Many 'boat people' drowned, were attacked and killed by pirates, or died from lack of food and water. Large numbers of unaccompanied children were on board these vessels, worried parents paying large sums of money to boat owners to try to get them to safety. If they were lucky, larger ships would intercept them and take them on board before dropping them off in countries all around south-east Asia.

Source 1 Vietnam and neighbouring countries

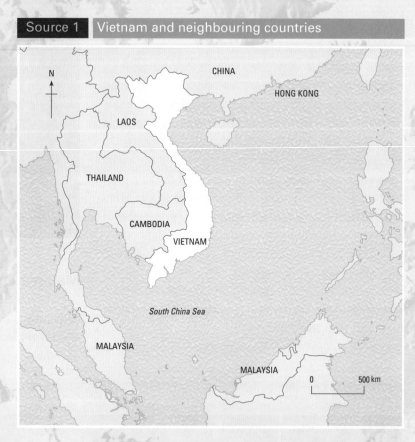

CHINA
HONG KONG
LAOS
THAILAND
CAMBODIA
VIETNAM
South China Sea
MALAYSIA
MALAYSIA
0   500 km

Source 2 Most boats carrying refugees were old, unseaworthy and very overcrowded

During the late 1970s, most of the boat people were genuine refugees, fleeing for their lives from persecution by the new communist regime. The majority arrived in Hong Kong, Thailand, Malaysia, the Philippines and Indonesia. Hong Kong saw some of the biggest numbers, locked into large camps (Source 3). This was meant to be a temporary arrangement, the arrivals being admitted as refugees and entitled to be resettled in third countries. Many early arrivals were resettled in the USA, Canada, the UK, France and Australia. The USA doubled their own quota to respond to the crisis that followed their own unsuccessful intervention in Vietnam.

By 1979, Hong Kong had over 350 000 boat people in refugee camps as other nearby countries began to refuse to take new arrivals. New agreements were made at the first UNHCR International Conference held the same year. These centred around the Orderly Departure Programme (ODP). This began well, and by 1980 half a million Vietnamese refugees had been resettled in new countries outside Asia. However, the situation quickly worsened as new arrivals again started to be refused entry. By now the majority were considered to be economic migrants, looking for work and a better quality of life away from Vietnam – many were from the north of the country, not the south.

In 1989, a second UNHCR conference introduced asylum screening processes as part of the Comprehensive Plan of Action (CPA). This aimed to sort out genuine refugees from economic migrants. Genuine refugees were still offered resettlement, whilst other migrants were encouraged to go home. Some agreed to be repatriated – but many refused to

| Source 3 | A Hong Kong detention camp: some detainees stayed for 20 years |

Fact File

- In 1975, US troops left Vietnam, Saigon fell and many south Vietnamese fled by boat.
- Over 2 million people were thought to have tried to escape from Vietnam.
- By 1996, 750 000 Vietnamese had been resettled; a third in western countries, the rest in Asia.
- 105 000 agreed to return to Vietnam.
- Many children were born in refugee camps in Hong Kong and Malaysia.
- By 1991, the number of arrivals dropped by over 85 per cent.
- Hundreds of thousands of Vietnamese families were – and still are – separated.

go and remained in refugee camps even after 1996 when UNHCR help ceased. Finally the last camp closed in Hong Kong in April 2000 when the remaining 1500 detainees were offered residency.

1 **a** Name the continent with:
  i the highest average birth rate
  ii the lowest average birth rate.

  b Suggest the reasons for the difference in the size of the birth rate between **a(i)** and **a(ii)**.

  c **i** Using Source 4 on page 119 to help you, explain why death rates have declined in most of the world.
  ii How has this decline affected world population change?

2 **a** **i** What is 'natural increase'? How is it calculated?
  ii What is 'migration balance'? How is it calculated?

  b Study Source 5 on page 119, the demographic transition model. Use the graph to help you describe how the relationship between birth and death rates changes over time.

  c **i** At what stage(s) are most LEDCs found?
  ii At what stage are most MEDCs found?
  iii Why are they at different stages?

3 **a** Briefly explain the differences between the following types of migration:
  i voluntary and compulsory
  ii internal and international
  iii permanent and temporary.

  b How might natural hazards lead to migration? Use an example you have studied to help illustrate your answer.

  c What are the main causes of both voluntary and compulsory migration?

4 **a** Explain the difference between 'pull' and 'push' factors which often influence migration.

  b **i** Give reasons for rural to urban migration in LEDCs.
  ii Give reasons for urban to rural migration in the UK.
  iii Why is rural to urban migration more important in Brazil?

  c 'The main direction of internal migration in both the UK and Brazil is from north to south, but the reasons for the two migrations are very different.' Explain this statement.

  d What problems are created by:
  i rural to urban migration?
  ii urban to rural migration?

5 **a** Why do so many Mexicans want to migrate to the USA?

  b **i** How does the USA classify different groups of migrants?
  ii Why is there such a high number of illegal immigrants from Mexico in the USA?

  c Use figures from Source 2, page 128 to draw two line graphs (using the same axis) for total legal migrants to USA and migrants from Mexico to the USA 1990–2002.

  d What problems are created by large-scale migration between the two countries?

  e How might maquiladoras:
  i encourage people to stay and work in Mexico
  ii make Mexicans want to migrate to the USA?

6 **a** Explain the differences between:
 i refugees and asylum-seekers
 ii internationally displaced persons (IDPs) and returnees.
 b Why are IDPs of such a concern to UNHCR?
 c The United Nations High Commission for Refugees is the major world organisation responsible for the welfare of refugees. Undertake your own research about UNHCR and write a brief case study using these headings: Formation, Structure, Aims.
 d Over 80 per cent of the world's refugees are found in LEDCs. Why? What problems does this create?

7 **a i** What caused the exodus of hundreds of thousands of Vietnamese from 1975?
 ii How did this change in the late 1980s?
 b i What dangers did these 'boat people' face?
 ii Why do you think there were so many unaccompanied children on the boats?
 c i Where did most of the boat people sail to?
 ii Where were they mainly resettled?
 d What impact did the Comprehensive Plan of Action (CPA) have on Vietnamese refugees and migrants?

Urban environments

Unit Contents

Chicago, home of the skyscraper. Which characteristics of a CBD are shown?

Urbanisation

Features of urbanisation

The growth of towns and cities which leads to an increasing proportion of a country's population living in urban areas is called **urbanisation**. Cities are growing in size all over the world. While the world's population is increasing fast, the urban population is increasing even faster. Source 1 shows that the world population more than doubled between 1950 and 2000 but that the urban population more than trebled.

What is significant about present-day rates of urbanisation is the difference in the speed of growth between the cities in the more economically developed countries and those in the less economically developed countries. The rate of city growth is much higher in the LEDCs (Source 2) so that the number of urban dwellers is now greater than in the MEDCs. Present trends are expected to continue.

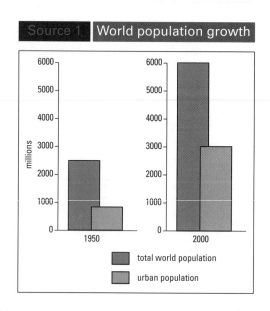

Source 1 — World population growth

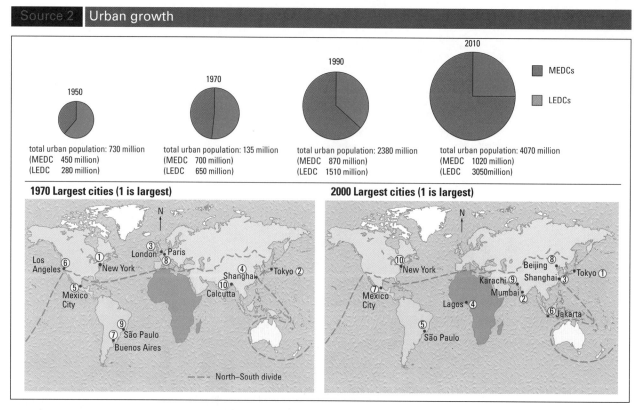

Source 2 — Urban growth

The maps for 1970 and 2000 show considerable changes in the world's 'top ten' cities. In 1970, half were in LEDCs and half in MEDCs. By the year 2000, only two of the ten were found in LEDCs. Not only has this changed dramatically, but so has the total number of people living in our largest cities. In 1970, the figures ranged from 16.5 million (New York) to 6.5 million (Calcutta). By 2000, the most populous city was Tokyo with 26.7 million people. New York, rated number ten in 2000, had fallen slightly to 16.3 million.

The size of big cities is another feature of world urbanisation. For many years the **millionaire city** (a city of more than one million people) was considered a big city, especially since in 1900 there were only two – London and Paris. Now there are about 400 (Source 3).

More recently, the term **megacity** has been used to describe cities with populations of over 10 million. In 1970 there were just four of these, but by 2000 there were 15. The United Nations estimate that by the year 2015 there will be at least 26 megacities, over half of which will be in Asia.

| Source 3 | Millionaire cities |

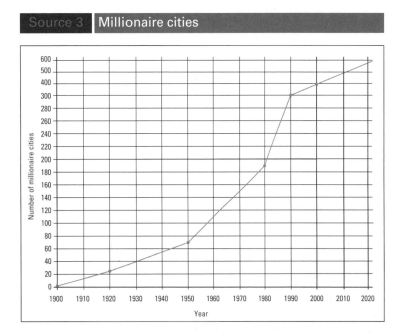

| Source 4 | Mexico City, the largest city in a LEDC |

Causes of urbanisation

Urban growth has always been associated with economic development. As a country increases in wealth, fewer people work in primary activities such as farming and forestry in the rural areas. Increasing numbers of people now work in secondary (manufacturing) and tertiary (service) occupations, which are overwhelmingly concentrated in urban areas.

High rates of urbanisation in LEDCs occur because:

- most new economic developments are concentrated in the big cities
- push and pull factors lead to high rates of rural to urban migration
- cities experience high rates of natural increase of population.

6.2 | Urban growth in a LEDC
São Paulo, Brazil

Source 1 | Aerial view of the central area of São Paulo

Source 2 | The location of São Paulo

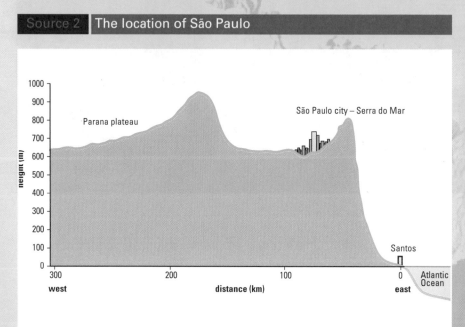

The built-up area of São Paulo in Brazil houses about 17 million people, making it the world's third largest urban area and one of its megacities. From the air it does not have the appearance of a city in the less economically developed world (Source 1). Its skyscrapers and modern concrete and glass buildings make it look more like New York. Yet way back in the 1870s it was described as a 'sleepy, shabby little town'. However, that was before the coffee boom in the interior of São Paulo state. The city's present size and appearance reflects its importance as the main industrial and business centre of Brazil. With a population of 165 million, there are more people in Brazil than there are in all the other South American countries added together.

The city is located on the plateau, not on the coast (Source 2). It was coffee growing on the Parana plateau inland from São Paulo which triggered its early growth. The flow chart in Source 3 illustrates how and why this happened. Profitability and growth led to more growth, a pattern which has continued to the present day.

During the 1950s and 1960s there was a burst of new economic growth in Brazil. Coffee remained important, but was left behind in the rush to industrialise. This industrial growth was driven by investments from multi-national companies, many with well-known names – VW and Ford, Philips and Sony. All chose to locate in São Paulo. Why? At that time São Paulo offered two main attractions as a location for manufacturing industry:

- Labour – an enterprising and hard-working labour force with some industrial skills.
- Energy – HEP, at first from stations at the base of the Serra do Mar, and later from dams along the big rivers on the plateau.

Success bred success. The pool of skilled industrial workers grew. It is continuously topped up by the inflow of migrant workers from rural areas such as the North-East; although they need to be trained, migrants tend to have the most drive and energy. University and research establishments in and around the city generate new ideas. The largest market of consumers in the country is on the doorstep. In this region is the highest density of paved roads and highways in Brazil, along which manufactured goods can be distributed to all parts of the country as well as to the port of Santos for export.

The positive result is that 40 per cent of Brazil's manufacturing output comes from São Paulo and surrounding areas. Although best known for its car, electrical and electronic industries, a full range of manufactured goods, from satellite dishes to canned drinks, is produced. The headquarters of most major companies are located in or on the edge of São Paulo's CBD, as also are those of the banks and other financial institutions.

The negative result is that the city continues to attract more migrants than it has jobs. City authorities have been unable to keep pace with the increasing demands for housing and for access to public services such as a clean water supply, sewerage removal, schools and health care. Air, water and ground pollution are serious problems. Traffic is gridlocked for most of the day, which has led to a boom in helicopter use as the favoured means of transport for rich business people.

Source 3	The early growth of São Paulo (1870–1950)

Coffee cultivation prospered on the fertile red soils of the Parana plateau.

↓

The coffee was sent to market in São Paulo.

↓

Out of their fabulous profits, coffee estate owners built mansions and invested in the city.

↓

Manufacturing industries to process the farm products were set up.

↓

Immigrants from Europe and Japan with their different skills were attracted.

↓

Coffee and manufacturing industries yielded big profits, some of which were reinvested.

Urban problems and solutions: LEDCs

The world's urban populations are growing rapidly, especially in LEDCs (pages 136–7). In São Paulo in Brazil, the population increased by 500 000 a year throughout the 1970s and 1980s – from 7 million in 1970 to 19 million in 2000. The reasons *why* people move to cities in LEDCs are dealt with Unit 5.3, pages 122–3. This rapid and often unplanned growth and development has created a range of problems, mainly because of the speed at which growth has occurred.

- **Housing**: Much of the rapid growth of LEDC cities has been caused by people moving in from rural areas or other parts of the country. When they arrive there is nowhere for them to live, especially as many are looking for cheap, low-cost housing. The majority of these new, usually poor, migrants have no option but to find a patch of land, either on the outskirts of the city or in areas no one else wants, such as steep slopes. Here they build a basic shelter from

Source 1 | Urban problems

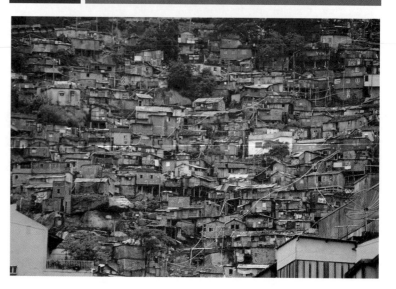

whatever materials can be found – often scraps of cardboard, wood and corrugated iron. Millions of people live in what were meant to be temporary **shanty towns** or **squatter settlements** in LEDCs. These areas are called favelas in Brazil (Source 1), bustees in India and barriadas in Peru.

Source 2 | Gridlocked traffic, Delhi, India

- **Access to water and electricity:** Whilst newcomers can build shelters for themselves and their families, few have access to running water, sanitation or electricity. Many have to light a fire to cook on, creating a major fire risk with shacks made from combustible material (Source 3.

- **Traffic congestion and transport:** The shanty areas themselves rarely have any organised public transport and few proper roads. The existing transport systems in the city are overloaded and overcrowded, and traffic congestion is a major problem for everyone – rich or poor (Source 2). The high numbers of vehicles also causes high levels of atmospheric pollution in cities, many of which suffer regularly from smog.

- **Health:** There are insufficient doctors, clinics or hospitals to deal with the rapid increase in population. With little or no access to clean water or sanitation, diseases and infections, such as typhoid and cholera, spread quickly. Atmospheric pollution leads to widespread respiratory problems.

- **Employment:** Although people are attracted to cities for work, many are unable to find proper paid work. Instead they are either unemployed or become part of the massive informal employment sector, surviving as best they can. This includes selling goods on the street (Source 4), working as cleaners or shoe-shiners or cooking and selling food from home or by the roadside. Even where there is paid work in new factories, these are often many kilometres from the shanty areas where most newcomers live.

- **Education:** Rapid population growth means lack of schools, although most manage some primary education. Few go on to secondary school because of the cost and because many children have to work to help support the family.

- **Social problems:** Given the close proximity and poor conditions experienced by sometimes millions of city dwellers, it is not surprising that they also suffer from high crime rates, drug trafficking and theft. Shanty towns are often inhabited by violent street gangs.

| Source 3 | Slum dwellings in Jakarta, Indonesia |

| Source 4 | Informal employment: a street vendor in the Caribbean |

Despite all these difficulties, many improvements have been made in LEDC cities, especially to some of the older shanty town areas (see pages 142–3).

Urban problems and solutions in a LEDC
São Paulo, Brazil

The growth of São Paulo has been described on pages 138–9. One of the world's largest **megacities**, its rapid growth (Source 1) and development as Brazil's largest and most industrialised city has created a number of problems, especially in terms of finding housing for new migrants. Like many growing cities in LEDCs, this has resulted in the growth of shanty towns, both around the edges of the city and within it.

Most people who migrate to São Paulo come from poor rural areas in search of work. There are no houses for them, so they build homes on the only land available usually on areas of no economic value, on the edge of town, along main roads or on steep slopes (Source 2) as part of a **favela** or **shanty town**. Many of these areas are prone to flooding or landslides or are in heavily polluted areas. However, for many, even living in a favela and working in the informal economy can be better and offer greater opportunities than the life they have left behind. About 20 per cent of São Paulo's population are migrants who fit this description.

São Paulo has approximately 2500 favelas. Some of the most well known are Helipolis (population 50–60 000), Paraisopolis (30 000) and Jaguare (24 000). All are away from the city centre but close to new factories. Some of the biggest favelas are up to 40 years old, and it is these which have seen the greatest level of improvement. When they were first built, few homes had even the most basic facilities, and there were no community facilities. Many people would illegally hook up to overhead electricity lines. Over time, communities developed and became organised. In Jaguare, a strong Neighbourhood Association has developed. By working together, people in Jaguare favela have persuaded the federal government to help them reduce crime and offer people, especially children, a wide range of sport or other activities.

Source 1 | The growth of São Paulo's population

(Graph: population (millions) vs year)

- 1970: 7.1 million
- 2000: 19.1 million
- 2015: estimated 23 million

Source 2 | A poor favela in São Paulo

Source 3 | An improved favela in São Paulo

Community groups have actively campaigned in many favelas to improve housing conditions and put pressure on city authorities to provide basic services like water, sanitation and electricity. Some of these are via self-help or site or service schemes. The government or an NGO provides building materials which local people use to build better homes (Source 3). Once built, water, sewerage and electricity services are provided. Rent has to be paid, so not everyone can afford this.

In some favelas, help has been made available so residents can get legal rights over their homes and land. This means that they are more secure, and can sell (and buy) property. With more stability this encourages further investment in favelas. There are also schemes to lend small sums of money to people running businesses in favelas via 'microlending.' Similar schemes have already been very successful in countries like Bangladesh. In Brazil, a **microcredit** scheme was launched in Heliopolis favela by a US non-profit-making organisation and a Brazilian bank working together. Local people with small businesses, such as bakers and grocery store owners, may apply for loans of between $100 and $1500.

However, as life improves in some favelas, there is concern that this will merely encourage more newcomers to the city. In the long term, there need to be other solutions. The best would probably be to

Source 4 | Edge cities like Bertine in Brazil may attract people away from existing centres

improve the quality of life in the rural areas of Brazil, but this would be very expensive. The largest cities like São Paulo are seeing new **edge cities** like Berrini and Jardines develop on their outer boundaries. These may help by encouraging migrants – and existing residents – away from the main city itself.

Urban land use and models

Urban land uses include shops, offices, factories, transport, recreation and waste land. However, the land use which covers the largest amount of land is housing. The term **morphology** is used to describe the layout of an urban area and the way in which the land uses are arranged within it. In most British cities it is possible to recognise three **urban zones** based upon location and land uses (Source 1).

Source 1	Urban zones in British towns	

Urban zone	Location and appearance	Land use characteristics
CBD (Central Business District) 	• city centre • tall buildings including skyscrapers containing offices • high building density with little open space	• old buildings, e.g. cathedral, castle • many shops of different types including department stores • company offices, banks and building societies • places of entertainment such as theatres and night clubs
Inner city (also known as the twilight zone or the zone of transition)	• around the edge of the city centre • unattractive, run-down appearance with many old buildings, made worse by vandalism and graffiti	• factories and warehouses • residential – often terraced houses and high-rise flats • universities and hospitals • inner ring roads • small shopping centres selling everyday convenience goods
Residential suburbs	• all the outer areas up to the edge of the built-up area • generally smarter appearance in the outer suburbs • some areas of open space	• residential – with the houses increasing in size, becoming more recent, changing from terraced to semi-detached and detached towards the outskirts • small shopping centres selling everyday convenience goods

Urban models

Various **urban models** have been devised to show the general arrangement of land use zones in cities (Source 2).

The Burgess model (Model 1) shows a circular pattern of land uses around the CBD. This model uses five zones because the inner city and residential suburbs have each been subdivided into two. The Hoyt model (Model 2) uses the circles of the Burgess model as its base but then adds sectors to show that similar land uses are concentrated in certain parts of the urban area. For example, factories may be concentrated in one area to form a zone of industry. A sector containing many high-priced houses may follow the line of a main road resulting in the formation of a high-class residential area for the wealthy. The Burgess and Hoyt models had to be adapted to show the general land use patterns in cities in the less economically developed countries. Model 3 includes the inner city slums and shanty towns which house many people.

There are three explanations for these land use patterns.

1 Historical

The urban area expanded outwards from the original site which is where the city centre is found today.

2 Economic

Rents and rates in the CBD became too expensive for people. In the suburbs there was more land and it was cheaper. Only businesses could afford to stay in the CBD, but even they needed to make the most of expensive land by building upwards.

3 Concentrations of similar land uses

One part of the urban area may have all the advantages for industrial location so that a lot of factories want to locate there; but few people want to live next door to a factory, so the residential areas are located elsewhere. Planners also prefer this **segregation of land uses** into definite zones.

Source 2 | Urban models for land use patterns

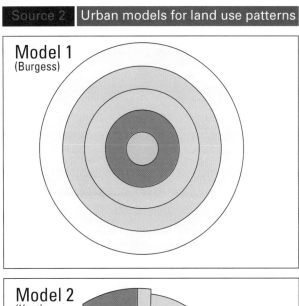

Model 1
(Burgess)

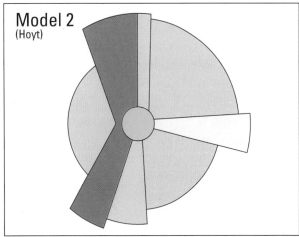

Model 2
(Hoyt)

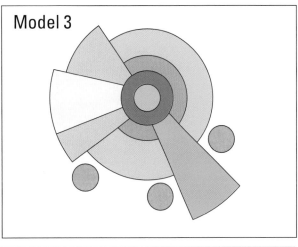

Model 3

Key

- Central Business District (CBD)
- industry
- inner city slums
- shanty towns
- light manufacturing
- old and low-class residential
- medium-class residential (inter-war housing in UK)
- high-class residential – modern housing

inner city

residential suburbs

Urban structure in a MEDC
Manchester, UK

In this unit we are going to follow a **transect** through Manchester to explore how land use changes. The transect runs north to south from Manchester city centre to the River Mersey along one of the main roads in and out of the city (Source 1).

Manchester's CBD shows up clearly on the map. It is where many main roads meet. The main railway and bus stations are found here, along with the city's main exhibition centre (G-MEX). The area is almost continuously built up with few open spaces.

The inner ring road which marks the southern boundary of Manchester's CBD is partly a motorway (in squares 8397 and 8497 on Source 1). The inner city begins to the south of this road. At first public buildings, particularly universities, hospitals and museums, take up a lot of the land, then residential land uses become more important from Moss Side and Rusholme southwards.

Source 3 shows data about the four Manchester wards you would pass through if you followed the transect from A to B (Source 1). Rusholme and Fallowfield are typical of residential areas found close to the city centre. Compared to the outer wards of Withington and Didsbury, houses are smaller, less than 50 per cent of households do not have a car, the majority are in lower paid jobs and unemployment considerably higher.

Source 2 | **Rusholme shopping centre**

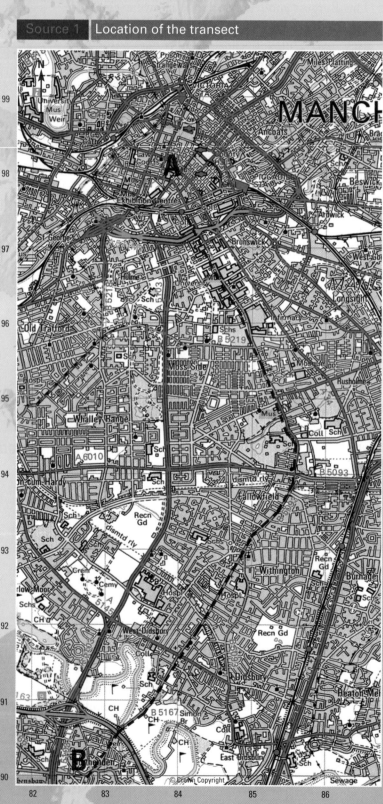

Source 1 | **Location of the transect**

Manchester is typical of many of the UK's older cities, with the quality of life and wealth of its residents generally increasing as you move away from the inner city area. Frequently newcomers and migrants to cities in LEDCs first settle fairly close to the CBD where housing is cheaper. In Manchester, there are much higher proportions of Asians and other ethnic groups in this area than elsewhere. Many migrants join established ethnic communities, continuing segregation.

In 1981, tensions spilled over into violence in Moss Side, another ward close to the city centre. Following a summer of unrest and riots in a number of UK cities, over 1000 mainly black youths surrounded the police station. With the country suffering from recession, unemployment in Moss Side was 44 per cent at the time and crime and drug-related offences had grown rapidly. Significant changes and developments have taken place since (see pages 154–7).

| Source 3 | Census information about Rusholme, Fallowfield, Withington and Didsbury |

Census information showing pie charts for RUSHOLME, FALLOWFIELD, WITHINGTON and DIDSBURY:

Types of housing (detached, semi-detached, terraced, flats)
- Rusholme: detached 2.3%, semi-detached 34.3%, terraced 34.3%, flats 29.2%
- Fallowfield: detached 5%, semi-detached 43%, terraced 24.9%, flats 27.1%
- Withington: detached 2.6%, semi-detached 48.8%, terraced 19.3%, flats 29.3%
- Didsbury: detached 8.2%, semi-detached 50.6%, terraced 15.3%, flats 25.9%

Ethnic group of the inhabitants (white, black, Asian, other)
- Rusholme: white 68.8%, black 7.3%, Asian 28.7%, other 4.2%
- Fallowfield: white 77.1%, black 6.7%, Asian 13%, other 3.2%
- Withington: white 87.2%, black 2%, Asian 8.5%, other 2.4%
- Didsbury: white 91.3%, black 1%, Asian 5.9%, other 1.9%

Households without a car
- Rusholme: 56%
- Fallowfield: 54%
- Withington: 42%
- Didsbury: 30%

Head of household working in a professional or managerial occupation
- Rusholme: 26%
- Fallowfield: 22%
- Withington: 32%
- Didsbury: 44%

Out of work
- Rusholme: 22%
- Fallowfield: 21%
- Withington: 13%
- Didsbury: 8%

147

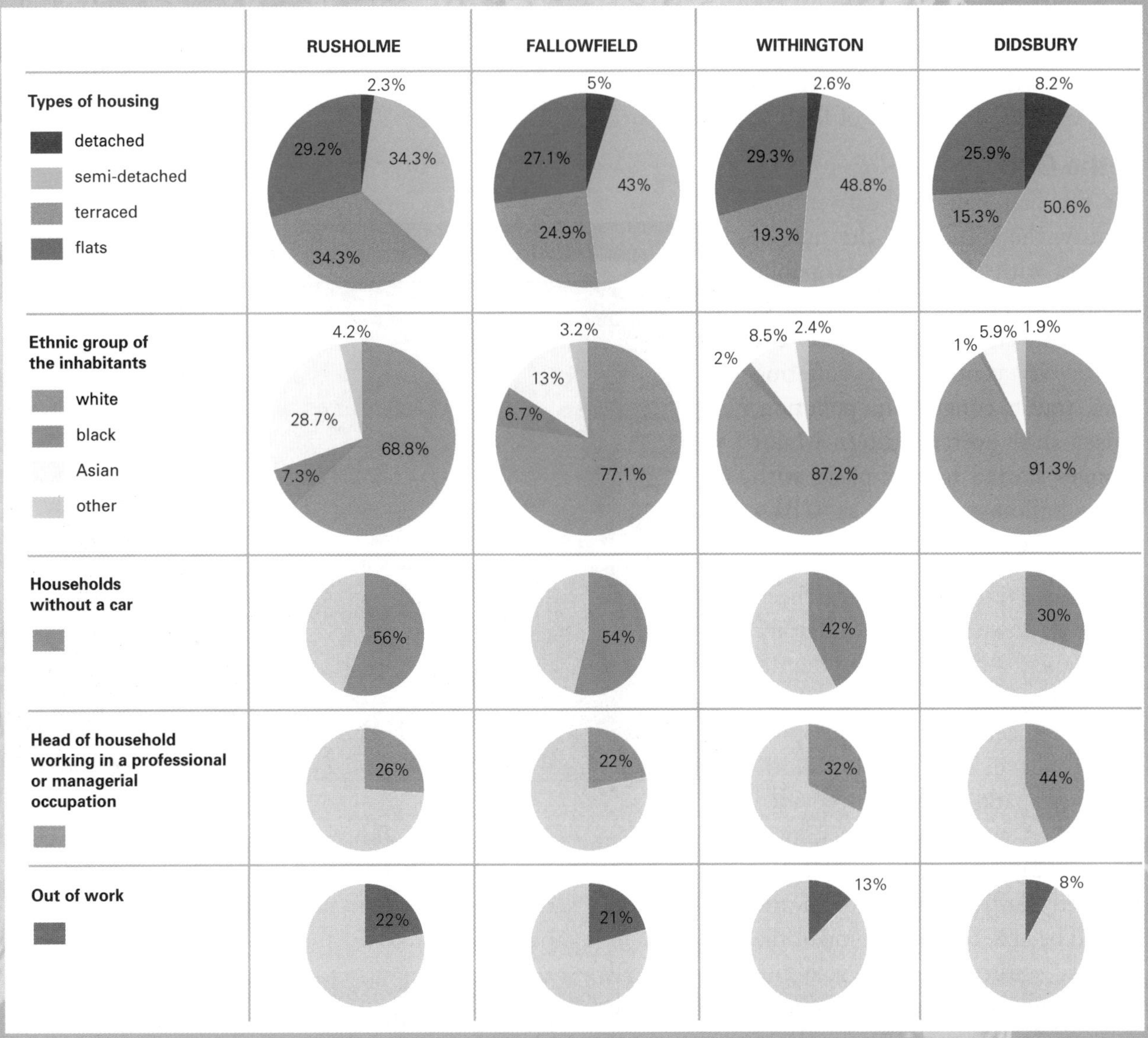

Cities are constantly changing, but whatever the changes are, whether they are growing rapidly or losing population, at the centre of most cities is an area called the Central Business District or CBD.

In many MEDCs, especially in Europe, towns and cities grew rapidly between 1750 and 1850 as industry grew rapidly during the Industrial Revolution. In the UK during this period, the population changed from 75 per cent rural to 75 per cent urban. Urban areas grew as people moved to work in new factories being built on what were then the edges of fairly small towns.

In the CBD

For the first half of the twentieth century, the CBD was the hub of the city with shops, offices, public buildings; it was where transport routes converged. By the end of the 1950s, city centres were suffering from traffic congestion, pollution, noise and overcrowding. Major changes started to take place, with many buildings cleared as CBDs began to be redeveloped.

Familiar old buildings have been knocked down to make room for skyscraper office developments. Main shopping streets have been pedestrianised. Indoor shopping centres have been built. Multi-storey car parks have been provided as part of city centre redevelopment schemes (Source 2). New one-way street systems have changed traffic flows through the centre. You should be able to notice some of these changes in the city centre nearest to your home. What all these changes are designed to do is to overcome city centre problems (Source 1). There is more urgency than ever to improve city centres because

| Source 1 | Changes in the city centre and the reasons for them |

Change	Reason for change
skyscraper office blocks	shortage and high price of land
inner ring roads	traffic congestion
one-way streets	traffic congestion
improvements in public transport: bus lanes, trams	too much use of the car for travelling to work and for shopping, parking problems
metro links between buses and cars, park and ride schemes	air pollution: photochemical smog and low-level ozone
pedestrianised shopping streets, indoor shopping centres	conflicts between shoppers and motorists

| Source 2 | A redeveloped city centre, Paris, France |

many of the functions of the traditional city centre are being threatened by out-of-town developments for shopping and business (see page 158).

Today, changes continue. If you visit almost any major city in the UK after a few years you would probably hardly recognise it.

In the inner city

Many inner city areas are depressing places in which change has usually meant decline and decay. City authorities and businesses have invested in the CBD. Much less has been spent in the inner city and here the environment is a growing problem. Large areas of waste land have become dumping grounds. Big factories are derelict monuments of the Industrial Revolution. Terraced houses, built for the better-off people in Victorian times, are now derelict and boarded up. Empty buildings are favourite targets for vandals and paint sprayers.

Economic decline follows. Those who could afford it have moved out, leaving behind the unemployed and the unskilled on low wages. Crime rates are high. The social character of the area has changed. Concentrations of people from the ethnic minorities are found in many inner cities. The numbers of pensioners, one-parent families and students are also above average.

The first wave of urban redevelopment in the 1960s and 1970s often saw old terraced houses bulldozed and replaced by multi-storey tower blocks, with people rehoused in flats. Thought to be a low-cost, high-density solution, within 30 years many were demolished. Source 3 shows some of the problems residents faced in the high rise buildings. Their construction also coincided with high levels of unemployment in the UK, especially amongst the young and ethnic groups, increasing social unrest in inner cities.

| Source 3 | Boarded-up and derelict inner city tower blocks |

| Source 4 | Experiences of living in high-rise flats |

walking up 15 floors when the lift is broken

using lifts which smell

fear of walking along the dark concrete balconies at night

no sense of community

worrying about children playing outside, ten floors below

nearby shops have closed down

Changing cities 2: planning, development and renewal

We know that our cities are changing all the time, but what are the reasons for these changes? Who is responsible for managing these changes?

Suburbanisation

As more people move to cities, cities need to grow to accommodate them. City centres grow upwards but most growth, particularly for housing, is outwards. This has created residential **suburbs** on the edges. Population density is lower than in the city centre, houses are usually larger and more expensive. Many people live in the suburbs and commute to the city to work.

Today outer suburbs are not just residential areas. From the 1970s in the UK, new out-of-town shopping areas were built on the edges of urban areas, taking advantage of increased space and cheaper prices (see pages 158–9). Now, many companies are moving their offices and factories to these locations, especially those with good transport links, such as motorways. In the USA, this has seen the building of **edge cities**, new cities in their own right built near the edge of existing large cities.

Counter-urbanisation

Since the late 1970s, many MEDCs have begun to experience counter-urbanisation, a decline in population as people moved away. At the same time, smaller towns and cities and rural areas have seen their populations increase. Source 2 shows some of the reasons for this change.

The widespread changes in city growth have heightened the debate over the type of land new

| Source 1 | Typical residential suburb in south-east England |

| Source 2 | The causes of counter-urbanisation |

- People wanting a better quality of life in quieter, cleaner rural surroundings
- More people willing and able to afford to travel further to work
- Relocation of companies (and jobs) to edges of cities or rural areas
- Flexible working practices and new technology – an increase in the ability to work from home for at least part of the week
- More retired people moving away from the cities where they once had worked

buildings and infrastructure should be built on. On the edges, new building is usually on 'greenfield' sites – land which has not been built on before, typically farmland. Many argue that this is damaging the countryside, especially when there are large areas of 'brownfield' sites available for building. This is land usually in urban areas which is derelict or not being used, such as old factory sites.

Planning change

Major changes in cities, whether new developments or renewal, need to be planned and managed. Typically large schemes are **public-funded** and led by national, regional and local government, often in partner-ship with other agencies such as housing associations. Private companies, land-owners, builders and devel-opers are also involved. Source 3 shows who may be involved in decision-making and managing change.

Large schemes to redevelop city centres or inner city areas take years to plan and build. Costs are typically hundreds of millions of pounds. Such projects are usually a mixture of facilities, for example shops, offices, industry, leisure, housing and transport, and are called **urban renewal** or **regeneration** schemes. Smaller developments also take place in cities. These are more likely to be funded by the **private sector**. It is quite common for large schemes to be a mixture of public and private sector funding – for example the redevelopment of London's Docklands in the 1990s (see Source 4). The UK government encouraged this type of partnership via Urban Development Corporations (UDCs) set up between 1980 and 2000. Manchester also set up an UDC to manage large-scale regeneration (see pages 154–7).

Whatever the project, government and landowners are the main 'players'. Government has the authority to approve developments and also has access to a wide range of public funding from within the UK and beyond. For example, grants from the EU have been used to set up enterprise zones to attract new industry to areas experiencing industrial decline.

| Source 3 | Managing change in UK cities |

Group	Main roles and responsibilities
Government/ politicians	Approve/disapprove schemes and applications; provide funding; set business and residential rates/taxes
Landowners	Sell land, set prices
Property developers	Find suitable land; buy from landowners; look for profitable developments and investors
Planners	Usually employed by local government; check building regulations; approve locations and building work
Industrialists/ business owners	Look for suitable sites for commercial property; availability of grants, workers, transport and access to markets
Other groups	For example: • housing associations – may help fund low-cost housing and rent out at affordable levels • pressure groups – may campaign against schemes impacting badly on the environment
Architects/ builders	Design and build approved developments in association with public/private sector
People	Represented by their elected politicians locally and nationally, but not necessarily directly empowered to prevent schemes they disapprove of

Landowners decide if they want to sell land, and how much it will cost – though in certain circumstances, government can compulsorily purchase it. In most UK cities, it is the local authority who has to approve and manage change, and deal with conflicting needs and demands.

| Source 4 | London Docklands: urban renewal |

The rural-urban fringe

The area around the edge of a city is known as the **rural-urban fringe**. It is where the green fields and open spaces of the countryside meet the built-up areas of the city. Countryside has been lost by the outward growth of cities and their suburbs. The open land around the edge of a city (greenfield sites) is in great demand for housing, industry, shopping, recreation and the needs of the public utilities, such as reservoirs and sewerage works.

One reason for growth and change in the rural-urban fringe is a feeling of dissatisfaction with the city.

- Houses are close together with few open spaces.
- Air quality is poor.
- Companies find that there is a shortage of land for building new offices and factories.

These are all **push factors**. There are also **pull factors** on the city edge.

- Land is cheaper so houses are larger.
- Factories can be more spacious and have plenty of room for workers to park their cars.
- Closeness to the main roads and motorways allows for quicker and easier customer contacts.
- New developments on the outskirts are favoured by the personal mobility allowed by the car.

Source 1 | Conflicts of interest between farmers and developers

People from the town trample on my crops.

I have lost half of my farmland to builders.

The public want to shop in out of town centres where parking is free and easy.

Businesses can make more profit if they are next to motorways.

My farm is split into two parts by the motorway.

Not everyone is happy with the continued loss of countryside around the cities. Many environmentalists are very concerned that new developments should be built on brownfield, not greenfield sites. There are often conflicts of interest, such as between farmers and developers (Source 1). People who have spent all their lives in villages resent the changes which they are faced with as villages grow into commuter settlements (Source 2).

Source 2 | Changes in a village as it becomes a commuter settlement

number of people in the village

300
250
200
150
100
50
0
1960 1970 1980 1990

average price of the old houses in the village

60 000
£
30 000
1960 1970 1980 1990

average number of cars passing through the village each day

160
120
80
40
1960 1970 1980 1990

change in morphology

new estate

old village

Key
■ old houses
□ new houses
— road

Retailing

In the more economically developed countries there has been a great increase in **out-of-town retailing**, with large purpose-built **superstores** and shopping centres located in the rural-urban fringe (see pages 158–9). The number of superstores has increased dramatically in the United Kingdom since 1980 (Source 3). It is easy to understand why. More people own their own cars. The large car parks are free. Access is easy because the shopping centres are located next to main roads and motorway junctions. In contrast, city centre shoppers face traffic congestion and expensive parking. Also the larger centres have shopping malls which are bright and modern with everything under one roof. Other facilities, such as multi-screen cinemas or bowling alleys, are often included within the shopping centre, or are located close by, so that there is something there for all the family (Source 4).

Source 3 | **Supermarket sweep**

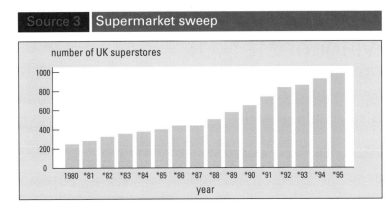

number of UK superstores

Source 4 | **Out-of-town shopping centre near Durham**

Source 5 | **Shopping hierarchy**

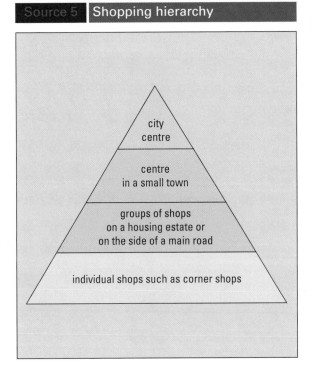

Out-of-town shopping centres do not fit the traditional **hierarchy** of shopping centres found in the UK (Source 5). At the bottom of the hierarchy is the corner shop, where everyday goods, such as milk, newspapers and sweets, known as **convenience goods**, are bought. These are **low-order goods**, used often, for which most people are prepared to travel only a short distance.

At the top of the hierarchy is the city centre. This is where people go to buy clothes or to have a look around department stores. Many of the goods sold in department stores, such as clothes, electrical items and furniture, are **comparison goods**. People buy these less often and are willing to travel further to buy them. The department store has a high **threshold**. A large number of people must shop there for the store to be profitable. It has a large **sphere of influence**: people travel from some distance away to shop there, helped by the fact that the city centre is the focus for the main roads and most bus and rail services.

6.10 Changing cities 4: a MEDC
Manchester, UK

History and growth

Manchester is one of the UK's major cities located in the centre of a major conurbation in north-west England (Source 1). With a population of 439 000, it is easily the largest of the ten districts which make up Greater Manchester, a metropolitan area created in the 1970s from two cities (Manchester and Salford), six towns (Bolton, Bury, Oldham, Stockport, Rochdale and Wigan) and two new boroughs (Trafford and Tameside). Greater Manchester's population is 2.6 million, but as many as 11 million live within just 50 miles of the city.

Although Manchester can trace its history back to Roman times, it was the Industrial Revolution which began in the seventeenth century that sparked rapid growth. By the late 1800s and early 1900s, Manchester had become one of the UK's most important industrial cities. It was the centre of the textile industry, with over 500 mills in Manchester and the surrounding county of Lancashire. These employed over 100 000 people producing cloth for the UK and overseas. Most of the workers moved to live next to the mills and factories, often in basic, overcrowded terraced houses in what is now the inner city, rapidly increasing the numbers living in the city (Source 2).

Other industries also grew rapidly in Manchester, many because of its importance as a port, linked to the sea by the famous Manchester Ship Canal. The area around the canal, Trafford, became a thriving industrial area with a wide range of industries – one of the UK's first large industrial estates. By the early 1900s, many well-known food processing companies had factories here, including CWS (the Co-op), Hovis, Kemp's, Kelloggs, Brooke Bond and over 300 USA companies, including Ford.

Source 1 | The location of Manchester

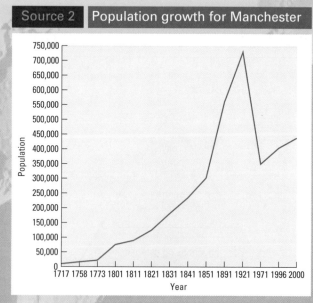

Source 2 | Population growth for Manchester

Inner city decline and change

The two world wars had far reaching effects on Manchester. During the First World War, it was impossible to export cloth to India, a major market. They turned instead to Japan and other suppliers. After war had ended, the trade continued to fall, drying up completely after India became independent in 1947. This was the beginning of the end for Manchester's textile industry, with the late 1950s seeing the closure of the mills and very high unemployment (Source 3).

The Second World War saw over 70 per cent of Manchester's older city centre buildings destroyed. Manchester played a major role in the war effort, its engineering skills used to manufacture weapons and machinery, building 1000 Lancaster bombers in just one year. This made the city a major target for enemy bombing. After the war, the City of Manchester Plan began, clearing the worst of the old terraced housing in the inner city. Large numbers of people moved out to the southern edge of Manchester, especially to the

| Source 3 | An old derelict cotton mill, Manchester, UK |

new 'garden suburb' of Wythenshaw which had started to be built just before the war. Most inhabitants of the inner city areas in the 1950s were new migrants to Manchester. Moss Side became the centre for Afro-Caribbeans, whilst many Asians made their home in Rusholme (see pages 146–7).

Between 1955 and 1975, 90 000 of the worst inner city houses were demolished to be replaced with a mixture of tower blocks and 'overspill' estates in inner city wards like Moss Side and Hulme (Source 4). These high-rise solutions introduced a range of new problems (see pages 146–7), and by the late 1980s and early 1990s, many were demolished. In the meantime, those who could move away did, and Manchester became another city experiencing counter-urbanisation.

In the city centre, old shops and narrow alleyways were replaced with the new Arndale Shopping Centre, a typical 1960s concrete-faced, yellow-tiled facility which many regarded as an eyesore.

continued

| Source 4 | New tower blocks were meant to solve Manchester's housing problems in the 1960s and 1970s |

Manchester, UK

Urban renewal and regeneration

Manchester in the 1960s–1980s was a city in decline. Manufacturing industry was unable to compete with competition from overseas and recession at home. In Trafford Park, north of the city centre on the Manchester Ship Canal, the number of companies operating had fallen from several thousand to just a few hundred. The docks finally closed in the 1970s and lay derelict. The area was designated an Enterprise Zone in the early 1980s, but little happened until it became an Urban Development Corporation in 1987. In the 11 years it was operational, it completely transformed Trafford Park.

The redevelopment via the UDC created opportunities for a large range of new businesses, residential and leisure facilities. This included major regeneration of the areas along the Ship Canal (Wharfside). Today the area is a major centre for finance and the arts in an attractive landscape setting along the waterside of Salford Quays. The Lowry Centre and footbridge (Source 5) is just one of the attractions in the area. To the west of the area, close to the M56, the Trafford shopping centre is one of the largest in the UK (see page 159).

Good transport structure and facilities have been a significant factor in Manchester's revival. Several major motorways link Manchester to the rest of the UK and Europe, and there is also a major Railfreight terminal in the city. Money has also been invested in Manchester's Metrolink system of linked trams and trains (Source 6). The first line opened in the 1990s running from Bury through Manchester to Altrincham. It has since been extended to Salford Quays and Eccles, with plans to link to Trafford Retail Park and Manchester Airport.

Source 5 Salford Quays: The Lowry Centre

Source 6 Metrolink map

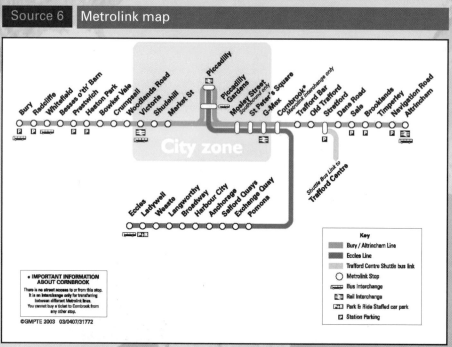

Reproduced by kind permission of GMPTE

In the early 1990s, Manchester was one of the cities which submitted a bid to hold the 2000 Olympic Games. Although the bid failed, the council had chosen a large area in east Manchester as the potential site for new stadia. The site was acquired and attracted initial £70 million of funding form both the local council and national government. Building began with the Velodrome – a new centre for UK cycling. A second, successful bid was made in 1995 to host the 2002 Commonwealth Games. This saw major investment and the building of the new City of Manchester stadium (Source 7), which later became the new home of Manchester City FC.

Manchester gained a great deal from holding the Games. East Manchester is now home to several nationally acclaimed sports venues, but has also benefited from new housing, community and business facilities. It has attracted major new investment to Manchester, created jobs and attracted an estimated additional third of a million visitors per year.

A variety of new housing has been built in the inner city. This includes the conversion of a number of former cotton mills and warehouses into luxury flats (Source 9), with prices up to £2 million each. This has attracted a new group of young professionals to live in the city. Moss Side and Hulme, two of the poorest inner city areas, have been regenerated since 2000 at a cost of £400 million, with improvements in place so that 'crime is designed out.'

Only two other English cities attract more visitors each year. After years of decline and stagnation, Manchester has been regenerated and is providing new jobs and homes as the population rises.

| Source 7 | City of Manchester Stadium |

| Source 8 | The Commonwealth Games legacy, from Manchester City Council's 'Lessons Learned' document |

'There is no doubt that the Commonwealth Games has helped place the area [east Manchester] at the forefront of regeneration initiatives both regionally and nationally. Without it, there would have been no unifying theme against which to justify bids for a wide range of regeneration programmes now running in the area.'

| Source 9 | Old warehouses redeveloped into luxury flats |

Although out-of-town supermarkets, superstores and small shopping centres are to be found around the edges of most large towns and cities in the UK, the big out-of-town centres are more narrowly distributed (Source 1). They are located close to very large centres of population, where there is good access, normally by motorway. The south-east of England is different in that there are two large centres close together, although they are situated on different sides of the River Thames. The presence of two reflects the size and wealth of the consumer market in London and the surrounding areas.

The success of out-of-town shopping centres has exceeded expectations. They are very popular with shoppers. Many are seeking planning permission to expand. Not everyone is in favour. Large out-of-town shopping centres are definitely not popular with local councils and high street shopkeepers from nearby towns, where shops close and 'For Sale' signs go up everywhere. The appearance of the high street deteriorates. Even fewer people want to go there. Council income from business rates goes down. A spiral of decline occurs. This happened in Dudley after Merry Hill opened. It only takes a few big names to abandon a town centre, such as Marks and Spencer and Sainsbury's, and many others follow. There are no longer sufficient shoppers for the larger shops to make a profit. Planners and environmental groups do not like the way greenfield sites are being swallowed up by out-of-town developments, which also encourage greater car use.

As a result, owners, developers and managers of the large centres are keen to drop the out-of-town label. They claim that they are becoming, or have already become, towns in their own right. At the MetroCentre people spend the same amount in a year as in a city the size of Oxford. The developers of Bluewater, opened in 1999, say that it may look like out-of-town in 2000; but with a planned 5000 houses to be built close by before 2010, they claim it will soon look like a new town.

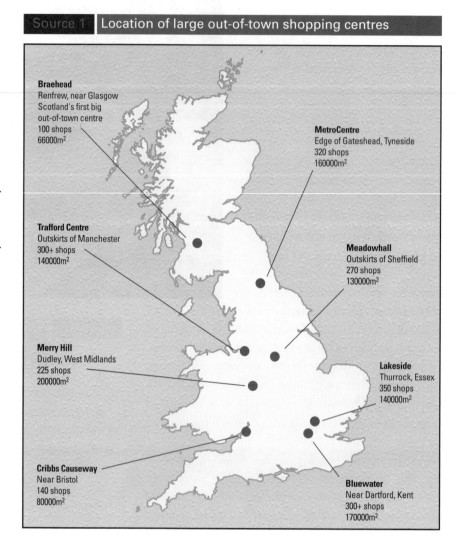

Source 1 Location of large out-of-town shopping centres

Braehead
Renfrew, near Glasgow
Scotland's first big
out-of-town centre
100 shops
66000m²

MetroCentre
Edge of Gateshead, Tyneside
320 shops
160000m²

Trafford Centre
Outskirts of Manchester
300+ shops
140000m²

Meadowhall
Outskirts of Sheffield
270 shops
130000m²

Merry Hill
Dudley, West Midlands
225 shops
200000m²

Lakeside
Thurrock, Essex
350 shops
140000m²

Cribbs Causeway
Near Bristol
140 shops
80000m²

Bluewater
Near Dartford, Kent
300+ shops
170000m²

Out-of-town shopping 6.12
Trafford Centre, Manchester, UK

The Trafford Centre is located on the north-west edge of Manchester, next to junction 4 on the M60. It opened in September 1998 and is the largest of its kind so far in Europe – a combination of shops, restaurants and leisure facilities. Built on the western edge of the Trafford Park area as part of the Manchester Urban Development Corporation's programme to regenerate the area, it occupies 300 acres of former wasteland in the Dumplington district. It took just over two years to build at a cost of £600 million. During its construction, it provided jobs for over 3000 building workers.

The new complex is easily recognised, especially at night when it is floodlit, with its three domes covering an area of 1.4 million square feet (the size of 30 football pitches). It attracts people from all over the north-west of England and further afield. Most come by car, but there are good local bus links and plans to extend Manchester's Metrolink rail/tram system to service the centre. An estimated 27 million visitors a year visit the centre. Like many similar developments, shopkeepers in nearby towns were concerned that they would see their own trade fall, possibly as much as 20 per cent.

As part of the Trafford Park redevelopment in Manchester, it is also close to the regenerated wharfside area along the Manchester Ship Canal, where the new Lowry Centre is located.

Fact File	The Trafford Centre

- Almost 300 major shops including Debenhams, Boots, Selfridges, Body Shop, Bhs and Marks and Spencer. A new John Lewis store is due to open in 2005
- 36 restaurants and fast-food outlets in an huge dining area called 'the Orient' which also has a 'Chinese Street', swimming pool and stage area
- 20-screen multiplex cinema – cinemas and restaurants remain open until midnight
- Covered market area at the end of Peel Avenue for independent/local traders
- Parking for 10 000 cars and 300 coaches
- Good facilities for disabled including a Shop-mobiilty scheme
- Children's play area, entertainment and crèche

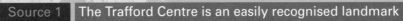
Source 1   The Trafford Centre is an easily recognised landmark

1 a i Using the data below draw a line graph to show the changes in the percentage of people living in urban areas in Brazil from 1940 to 2000.

Year	Percentage living in urban areas
1940	31
1950	36
1960	45
1970	55
1980	66
1990	76
2000	81

    ii Describe what the data shows.
b i What is meant by urbanisation?
    ii State three causes of urbanisation.
c i What is a megacity?
    ii Describe and explain how the numbers of megacities is changing.
d Give as many reasons as you can why cities in LEDCs are growing faster than cities in MEDCs.

2 a Describe the main features of the central area of São Paulo shown in Source 1 on page 138.
b i List São Paulo's manufacturing (secondary) industries.
    ii Describe three factors that explain the huge growth in manufacturing (secondary) industries in São Paulo.
c What problems have been caused to the city by rapid industrial growth?

3 Choose a large city an LEDC. For your named city:
a Describe:
    i the main urban problems apart from housing
    ii the main causes of these problems.
b Explain why finding solutions to these problems are difficult.
c Describe:
    i housing problems in your named city
    ii what is being done to try to improve housing.
d How does housing and quality of life change as you move away from the city centre to the suburbs? You may want to use a named city in your answer.

4 a i Why is the city centre called the Central Business District (CBD)?
    ii Why are some of the highest buildings found in the CBD?
    iii Make a list of the main buildings/land use found in the CBD.
b i Describe how Burgess' Land Use Model (Model 1) has been adapted and changed for Hoyt (Model 2) and Model 3.
    ii What are the advantages and disadvantages of land use models?

5 a Much inner city redevelopment in the mid twentieth century, especially housing in the UK, is often thought of as a 'planning disaster'.
    i Describe the type of changes which took place.
    ii Explain why they were soon thought of as 'disasters'.
b For a city in an MEDC you have studied, for example Manchester, describe:
    i how the inner city area has been redeveloped
    ii the reasons for redevelopment.

6 a Explain what is meant by:
    i suburbanisation
    ii counter urbanisation.
b What is an edge city?
c **i** List the different people or groups who have an interest in or influence on city planning.
    ii Rank your list in order with those who has the most influence at the top of the list and those with the least at the bottom.

7 a **i** What is a UDC?
    ii Why were UDCs created?
b Trafford Park to the north of Manchester along the Manchester Ship Canal has been an important industrial area for over 100 years.
    i What was the area like from 1900 until 1950?
    ii Why did it decline in the 1950s and 1960s?
    iii What changes have taken place in recent years?

8 a **i** Describe what we mean by the 'rural-urban fringe'.
    ii What are the main push and pull factors in this type of area?
    iii Describe some of the conflicts that might take place.
b What is the difference between greenfield and brownfield sites?
c Explain why this type of location is attractive for out-of-town shopping centres.
d Choose one out-of-town shopping centre in the UK. Write a case study using the following headings: Site and situation, How the centre is reached, Shops and other facilities.

9 This unit looked at two major cities in detail – São Paulo in Brazil and Manchester in the UK. Write a brief account of:
- how they are similar
- how they are different.

Fragile environments

Unit Contents

- Fragile environments and sustainability

- Soil erosion

- Desertification

- Desertification: the Sahel, Africa

- Deforestation

- Deforestation: Amazon rainforest, Brazil

- The greenhouse effect and global warming

- Coping with climate change

A typical landscape in the Sahel. How and why does the land look like this?

Fragile environments and sustainability

Source 1 World biomes

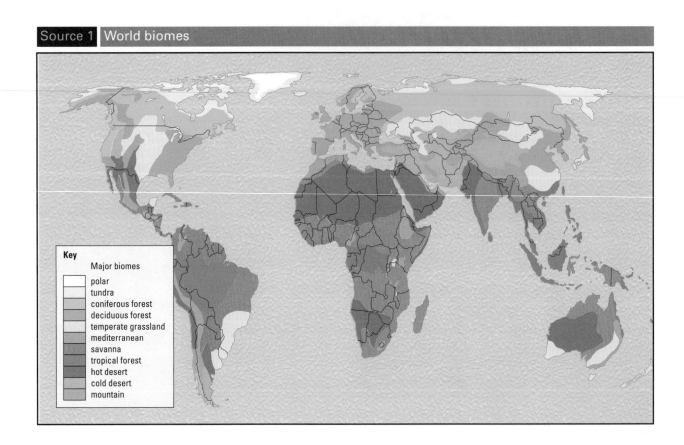

Key

Major biomes
- polar
- tundra
- coniferous forest
- deciduous forest
- temperate grassland
- mediterranean
- savanna
- tropical forest
- hot desert
- cold desert
- mountain

The earth contains a range of unique environments called **biomes** or **ecosystems** (Source 1), each with specific characteristics. These depend on a specific combination of living and non-living parts including climate, rocks, soils, natural vegetation, animals and human activity. People have always made use of ecosystems to provide themselves with food, fuel and building materials. Even today, there are a few indigenous tribespeople like the Baka in Africa who live in harmony with their local environment, using it in a sustainable way.

Fragile environments are those biomes or ecosystems under threat from change, damage or unsustainable use – usually from human activity. Although natural hazards such as earthquakes, volcanoes and hurricanes can cause major damage, it is the action of people that has led to widespread destruction of natural environments. Other 'natural' hazards like drought and flooding have often been made worse by human action.

As land becomes more and more precious, the pressure on the land which remains undeveloped increases. We want to conserve and protect the **biodiversity** of plants and animals which live on our planet, but in our eagerness to visit and see this we are in danger of destroying it, especially the most fragile and rare elements.

Large areas of land on the edges of existing deserts are turning into deserts themselves. Overgrazing, the removal of trees, **soil erosion** and decreasing rainfall are all to blame, as land which was once useful to farmers becomes useless. The **exploitation** of natural resources in the world's rainforests leads to **deforestation** and destruction.

Source 2	Headline news – environments under threat

GALAPAGOS OIL CATASTROPHE
Leaking oil from stricken tanker threatens unique wildlife

DEATH SENTENCE FOR THE AMAZON
$40 billion project set to destroy 95% of rainforest by 2020

TREELESS HILLS SEND TORRENTS INTO INDIA
More than 100 killed and 2 million left homeless by floods and landslides in NE India

ANTARCTIC FISH IN TROUBLED WATER
Ozone depletion could destroy one of the world's last great stocks of fish

1000 FLEE AS SEA BEGINS TO SWALLOW UP PACIFIC ISLANDS
Inhabitants on coral atolls off Papua New Guinea are the latest victims of rising sea levels

Human and industrial waste pours into our rivers, seas and oceans. This may be accidental, or deliberate. Oil spillages from tanker accidents, and the dumping of sewage, toxic chemicals and other waste cause widespread pollution.

Many problems which appear to be local often contribute to more widespread problems. Traffic in towns causes congestion and pollution. Building new roads to solve these problems creates others, for example the destruction of rural environments, and increases in traffic. This in turn can lead to the formation of acid rain, the production of **greenhouse gases** and, ultimately, **global warming**.

If the diversity of life and environments is to survive, careful management is necessary. Pollution and conservation know no national boundaries. Decisions made at local level often have far more wide-reaching effects. International co-operation and legislation may be the only solution if people and countries are to behave responsibly to protect and sustain the earth for future generations.

Source 3 lists some of the ways in which the earth is being used in an unsustainable way. To achieve

Source 3	Examples of unsustainable global development

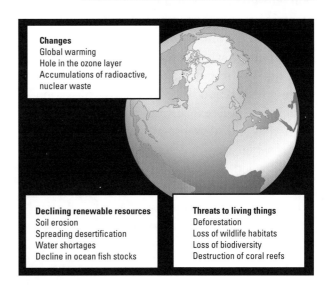

Changes
Global warming
Hole in the ozone layer
Accumulations of radioactive, nuclear waste

Declining renewable resources
Soil erosion
Spreading desertification
Water shortages
Decline in ocean fish stocks

Threats to living things
Deforestation
Loss of wildlife habitats
Loss of biodiversity
Destruction of coral reefs

sustainability, you need to fulfil your own needs without causing irreparable damage to the environment. For example, when trees are cut down for timber, they should be replaced, with due consideration of the time it takes for trees to mature. If resources can be managed sustainably, present and future demands for food, shelter, clothing and recreation will be met.

Soil erosion

There are three main types of soil erosion. In parts of the world where there is enough rainfall, exposed soil will be removed down slopes as a mass movement – **sheet erosion**. Source 1 shows a scar of sheet erosion on a hill slope in eastern Brazil caused by removal of the forest to create a coffee and orange plantation.

In some places where rain falls as heavy storms between drier times, soil will be removed in great gulleys – **gulley erosion** (Source 2).

In dry parts of the world, often on the edges of deserts, the loose, dry soil will be blown away by the wind – **wind erosion** (Source 3).

As populations increase in both LEDCs and MEDCs, so the pressure on the land increases. If there is too much pressure the land suffers.

Soil erosion can be caused by the physical processes outlined above, or by human action. The most common human causes are rapid population growth, increasing pressure on the land and its resources; overgrazing the land, causing vegetation loss; intensive cultivation leading to soil exhaustion and the loss of nutrients and deforest-ation. Deforestation occurs when trees are cut down for fuel wood or to clear land for farming (see pages 170–1). In practice, soil erosion is usually caused by a combination of factors, both physical and human.

| Source 1 | Sheet soil erosion |

| Source 2 | Gulley soil erosion in norther Tanzania |

| Source 3 | Wind soil erosion |

Desertification 7.3

Source 1 | The world's hot deserts and areas in danger from desertification

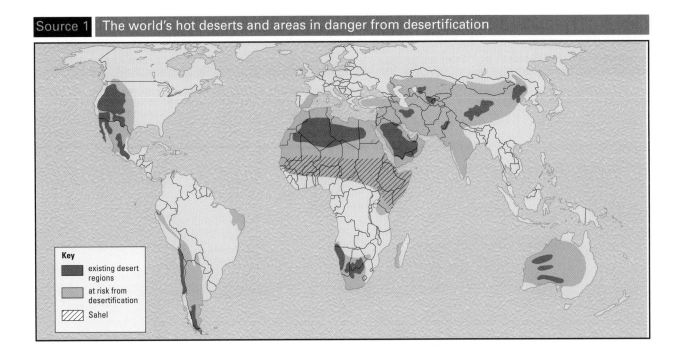

Key
- existing desert regions
- at risk from desertification
- Sahel

Desertification is the term used to describe how once productive land changes into a desert-like landscape. The process is not necessarily irreversible and, as Source 1 shows, usually takes place in semi-arid marginal land on the edges of existing hot deserts. As with soil erosion, there are both physical and human causes of desertification.

Physical causes include:

- soil erosion – exposed soil is easily removed by wind or water
- changing rainfall patterns – rainfall has become less predictable over the past 50 years and the occasional drought year has sometimes extended to several years
- intensity of rainfall – when rain does fall it is often for very short, intense periods. This makes it difficult to capture and store and can increase soil erosion.

The main human causes are:

- population growth – rapid increase in population puts more pressure on the land as more food is needed

- overgrazing – too many goats, sheep and cattle can destroy vegetation
- overcultivation – intensive use of marginal land becomes exhausted and crops will not grow
- deforestation – trees are cut down for fuel, fencing and housing. The roots no longer bind the soil, leading to soil erosion.
- war – many sub-Saharan countries have suffered from years of civil war. Crops and animals have been deliberately destroyed resulting in famine and widespread deaths.

As with many other threatened fragile environments, it is usually a combination of different causes that results in damage to a specific location.

It is estimated that about 20 per cent of the world's population have to cope with the effects of desertification in over 60 countries. One of the regions most at risk is in the Sahel region of Africa – an area south of the Sahara desert stretching the width of the continent. It forms a large part of Sub-Saharan Africa, the poorest region of the world.

| Source 1 | Location of the Sahel |

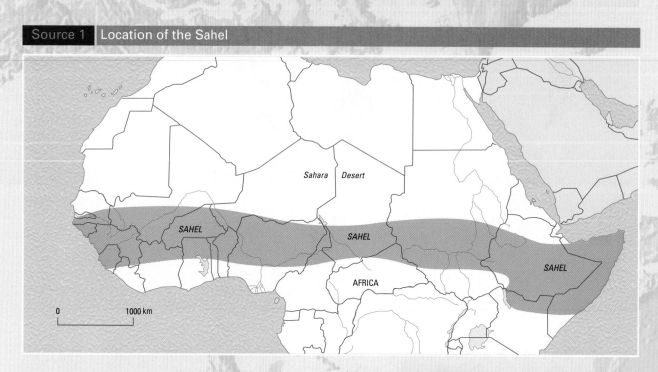

The Sahel is a narrow belt of land in North Africa. It borders the southern edge of the Sahara Desert (Source 1). The Sahel has a semi-desert climate shown in Source 2. Temperatures are always hot and there is a long dry season from June through to January. There is just enough rainfall for grasses to grow as well as some shrubs and trees.

The world biomes map on page 167 shows that the Sahel region is in an area classified as tropical savanna. The natural vegetation of tropical savanna is a mixture of grassland, trees and shrubs. However, the amount of each changes as you move away from the edge of the tropical rainforests and the Equator, where it is wetter and vegetation is much richer, to thinner grassland with occasional shrubs and trees where tropical savanna meets hot desert. It is these areas which are amongst those most at threat from desertification.

On the equatorial edges of savanna there are also more animals. As the trees thin out towards the middle of savanna regions, large herds of wild animals like wildebeest, antelope and zebra are found. On the drier desert edges like the Sahel region there is far less wildlife. Rainfall is seasonal and unpredictable and it is very dry for much of the year. Here you often find nomadic herders who move from place to place with goats and cattle in search of water and grazing.

| Source 2 | Climate graph of the Sahel |

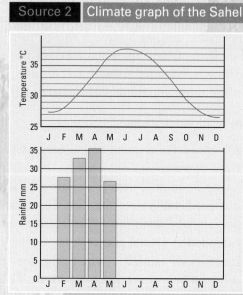

In the Sahel there have been some years when less rain has fallen. Fewer grasses have grown and trees have died. The landscape becomes much more like desert (Sources 3 and 4).

This climate change is one cause of **desertification**. Desertification is also speeded up by human activity. Up until the 1960s there was more rainfall in the Sahel and the population grew. Water was plentiful so more crops were grown and numbers of animals increased. Wood was available to use as fuel and building materials.

In the drier years after the 1960s large areas of forest were removed for farmland and fuel supplies. People still tried to grow the same crops and rear the same numbers of animals. **Overgrazing** and **overcultivation** left the ground bare. Without vegetation less humus is added to the soil. The soil holds less water and dries out. The bare soil is quickly eroded by wind and flash floods. The land can no longer support any trees and plants and it turns to desert. Since 1970:

• there has been widespread crop failure
• over 100 000 people and millions of animals have died.

A variety of techniques can be used to prevent desertification and also to rehabilitate the land that has already been damaged.

One successful method of catching rain when it falls is a simple technique set up by Oxfam in Burkina Faso. Small stone walls are built following the slope of the land which then act like dams when the rain falls, stopping surface run-off and allowing it to sink into the soil. This simple, inexpensive method can increase yields by up to 50 per cent.

New research using satellite images suggests that the deserts may not be spreading permanently. Some areas where rainfall has increased have now recovered. It is also difficult to decide whether it is the climate or the human activity which causes the changes. One thing is certain – the semi-arid lands are fragile environments and people must use them with care so that desertification is avoided.

Source 3 The Sahel before desertification

Source 4 The Sahel after desertification

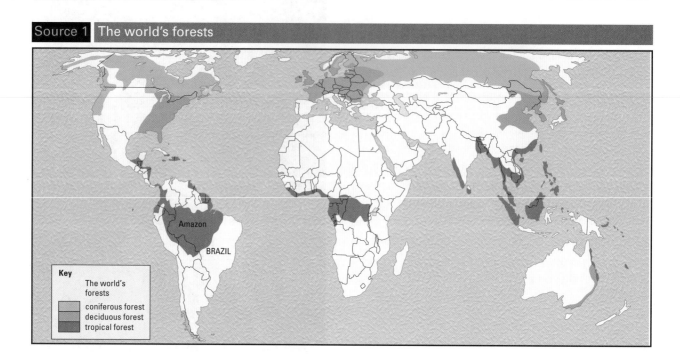

Source 1   The world's forests

Key
The world's forests
coniferous forest
deciduous forest
tropical forest

Deforestation is the cutting down of trees. Many primary forests in temperate countries have almost disappeared after centuries of logging or land clearance for farming, industry and housing. However, it is only fairly recently that large-scale deforestation has started to take place in the world's tropical rainforests (Source 1). The speed of deforestation has alarmed scientists and conservationists and the future welfare of tropical rainforests is an important environmental issue.

Areas of tropical rainforests are cleared for a variety of reasons, including logging, farming, road building, mining and HEP schemes. Sometimes this has been illegal or uncontrolled, but sometimes governments have encouraged the clearing of the forests because:

- the revenue earned from selling timber, drugs and minerals helps to pay off debts and to develop their countries
- new land is needed to house and feed the growing populations in countries such as Brazil and Malaysia.

One of the major concerns today is not just for the loss of biodiversity within each area of lost rainforests, but the global impact of deforestation. The rapid build-up of greenhouse gases in the atmosphere has started to increase global temperature. One of the most important greenhouse gases is carbon dioxide. The world's forests use up CO_2 from the atmosphere – by cutting them down more CO_2 remains. If the trees which are cut are then burnt, it releases even more CO_2.

The management of forests

The loss of forests not only in the tropics but all over the world is causing concern. Many governments and international organisations recognise the need to manage forests to ensure the resources are there for future generations. This is called sustainable development. But it is difficult for some governments, especially of the LEDCs, because they need the money the developments bring.

The sustainable management of forests can be achieved by:

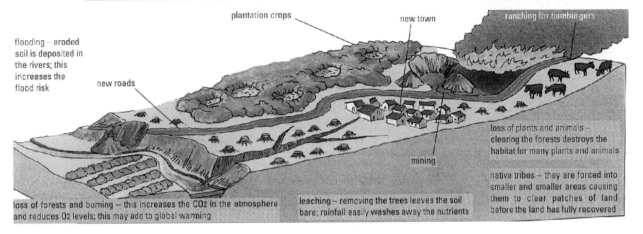

Source 2 Problems caused by deforestation: the problems affect the soil, vegetation, climate, rivers and local people – find an example for each one

plantation crops

new town

ranching for hamburgers

flooding – eroded soil is deposited in the rivers; this increases the flood risk

new roads

loss of plants and animals – clearing the forests destroys the habitat for many plants and animals

mining

native tribes – they are forced into smaller and smaller areas causing them to clear patches of land before the land has fully recovered

loss of forests and burning – this increases the CO2 in the atmosphere and reduces O2 levels; this may add to global warming

leaching – removing the trees leaves the soil bare; rainfall easily washes away the nutrients

- protection of forests – in some countries areas of forest are conserved and protected as National Parks where no development is allowed to take place
- carefully planned and controlled logging in forests
- selective logging of only those trees that are valuable leaving the rest of the forest untouched, for example in parts of Indonesia only 7 to 12 trees per hectare are allowed to be felled
- replanting of forested areas that have been felled
- restrictions on the number of logging licences that are allowed to reduce the amount of forest loss

- heli-logging, for example in Sarawak where helicopters are used to remove the logs because less damage is done to the remaining forest
- developing alternative energy supplies, for example biogas, solar and wind power to reduce the amount of wood needed for fuel.

What is sustainable development?

Today, people are supporting the idea of sustainable development. This is the ability of one generation to hand over to the next at least the same amount of resources it started with. It should also be development which helps all people, particularly the poorest. Sustainable development should:

Source 3 Logging in Sarawak

- respect the environment and cultures
- use traditional skills and knowledge
- give people control over their land and lives
- use appropriate technology – machines and equipment that are cheap, easy to use and do not harm the environment
- generate income for communities
- protect biodiversity.

Deforestation
Amazon rainforest, Brazil

Brazil is the largest country in South America. In the north, the climate is equatorial. The hot, wet greenhouse conditions produce very rapid growth of vegetation all year round. The tropical rainforest of Brazil's Amazon region is the largest in the world. Study Source 1 and notice the five layers of vegetation and the large variety of different trees and plants. There are over 1000 different tree species, for example mahogany, teak and rosewood.

For centuries the Amazon rainforest has been inhabited by groups of Indians. The Indians hunt for animals, collect fruits and clear small patches of the forest to grow crops. This type of farming is called slash and burn. The clearing is small and after two or three years it is abandoned and the forest once again develops. The Indians do no long-term damage to the forest and their use of the forest's resources is sustainable.

Recent human activity in the rainforest has been much more devastating. Large-scale deforestation has taken place for a variety of reasons:

- mining, for example the iron ore mine at Carajas in the Amazon Basin
- road building, for example the Trans-Amazonian Highway in Brazil
- new settlement and small farms to house migrants from the cities
- logging for timber exports, for example teak and mahogany in the Amazon Basin
- huge cattle ranches
- other types of farming, including plantations
- reservoirs and dams for a number of hydroelectric power schemes, for example Itaipu, Tucurui and the Xingu complex.

Source 1 | Tropical rainforest

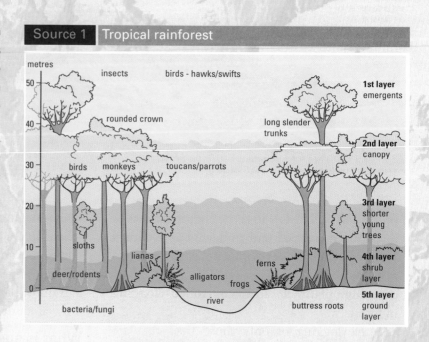

Source 2 | Buttress roots

The huge buttress roots at the base of the trees help to give the trees support and to take in the vast quantities of water and nutrients the trees need to survive.

There have been many consequences or effects of deforestation in the Amazon rainforest:

- Mining is typically open-cast, and destroys large areas of vegetation and pollutes rivers. Mercury is especially toxic.
- Roads have opened up the Amazon, allowing easier access and encouraging uncontrolled development in addition to planned, new settlements.
- When loggers fell the more profitable hardwood trees like teak, ebony and mahogany, they also unnecessarily destroy large areas of the forest.
- Clearing land for farming e.g. pasture for cattle not only releases CO_2 as trees are burned, but leads to nutrient loss and soil erosion. It is the rainforest vegetation which provides the fertility for otherwise poor soils.
- Large areas have been flooded and indigenous tribes moved from their lands to create HEP schemes.
- Biodiversity has been lost as habitats for animals, birds and insects are destroyed, alongside a wide variety of medicinal plants.

Source 3 The effects of deforestation in the Amazon

Source 4 The Trans-Amazonian highway

173

The greenhouse effect and global warming

Some scientists are not convinced that the global warming taking place today is largely the result of human activity. They hold the view that recent changes are part of the pattern of variations in world climate and/or could be connected with the activities of the sun. However, the majority view is that the rising temperatures and global climate changes taking place today are mainly due to the release of **greenhouse gases**, mostly via human activity. Source 1 lists the main greenhouse gases and how they are released into the atmosphere.

Source 1	Greenhouse gases

Gas	Causes
Carbon dioxide (CO_2)	Release of carbon from fossil fuels in power stations and through vehicle exhausts. Burning of wood. Deforestation – trees use up CO_2 from the atmosphere; without them CO_2 remains.
Methane (CH_4)	Decay of organic matter – waste in landfill sites, animal manure, large areas of crops, e.g. rice.
Nitrous oxides (NO_2)	Burning fossil fuels – car exhausts and power stations. Use of fertilisers.
CFCs	Gas released via aerosols, coolants in fridges, freezers and air conditioning systems, certain packaging and insulation.

Greenhouse gases occur naturally in the atmosphere. However, they can also be produced and released by farming and industry. Carbon dioxide is the main cause for concern as it makes up over two-thirds of the current level of greenhouse gases. Factory emissions, burning fossil fuels like oil, coal and gas in power stations and exhaust emissions from motor vehicles are the major sources of 'extra' emissions. The main producers are MEDCs, with the USA alone responsible for 36 per cent of all greenhouse gas emissions. However, the effects are felt by everyone.

Source 2	The greenhouse effect

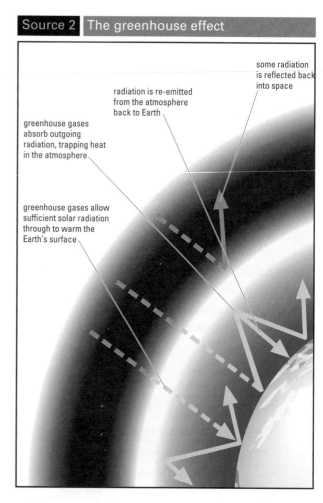

some radiation is reflected back into space

radiation is re-emitted from the atmosphere back to Earth

greenhouse gases absorb outgoing radiation, trapping heat in the atmosphere

greenhouse gases allow sufficient solar radiation through to warm the Earth's surface

What do we mean by the 'greenhouse effect'?

In a greenhouse, sun shines through the glass warming up the plants inside. When the sun stops shining, the heat does not disappear, it is trapped inside the greenhouse. In the same way, heat is trapped in the earth's atmosphere. During the day, radiation from the sun heats the earth. At night, clouds often trap this heat as it radiates back out. Gases in the atmosphere also trap this heat. This is the **greenhouse effect**. In recent years, the amount of these greenhouse gases has greatly increased. The main gases and their causes are shown in Source 1.

Source 3 | Possible effects of global warming

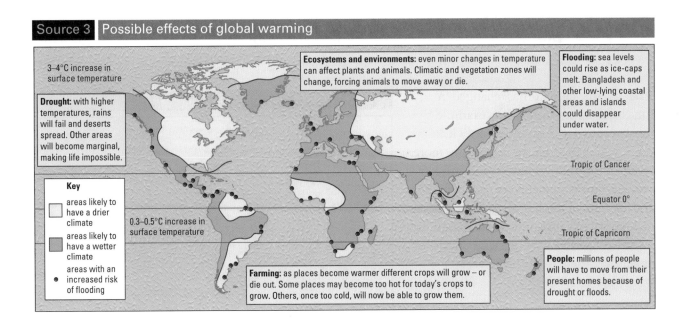

3–4°C increase in surface temperature

Drought: with higher temperatures, rains will fail and deserts spread. Other areas will become marginal, making life impossible.

Ecosystems and environments: even minor changes in temperature can affect plants and animals. Climatic and vegetation zones will change, forcing animals to move away or die.

Flooding: sea levels could rise as ice-caps melt. Bangladesh and other low-lying coastal areas and islands could disappear under water.

Tropic of Cancer

Equator 0°

Key

- areas likely to have a drier climate
- areas likely to have a wetter climate
- areas with an increased risk of flooding

0.3–0.5°C increase in surface temperature

Tropic of Capricorn

People: millions of people will have to move from their present homes because of drought or floods.

Farming: as places become warmer different crops will grow – or die out. Some places may become too hot for today's crops to grow. Others, once too cold, will now be able to grow them.

Global warming

Greenhouse gases build up in the atmosphere, preventing heat from radiating back out. This build-up is the main cause of the gradual increase in world temperatures known as global warming. Measurements over the past 100 years have shown an average rise of 0.7°C. The 1990s saw some of the highest temperatures ever recorded. If the rise in CO_2 levels continues to increase at current rates, it could lead to global temperature rises of between 2 and 5 °C over the next 50 years.

Source 3 shows some of the possible effects of global warming. Whilst higher temperatures could lead to major changes in farming, increased drought and changes to existing ecosystems, it is the effects of sea level rises which may have the greatest impacts. Higher temperatures have already led to the shrinking of many of the worlds major glaciers. In Greenland and the Arctic, ice is melting at a rapid rate and the extent of Arctic pack-ice is shrinking. In Antarctica, especially around the peninsula, sections of ice sheets are breaking off as temperatures rise. With so much ice and snow locked up in polar regions, global warming causing melting could lead to widespread flooding of the world's low lying islands and coastal areas, with disastrous consequences.

Source 4 | Melting polar ice could lead to rises in global sea level

Coping with climate change

The Earth's climate has changed many times during its history, for example the most recent Ice Age ended only 12 000 years ago. However, it is only in more recent times that we have made and kept accurate measurements of our weather and climate. This has allowed scientists to identify changes and trends based on recorded evidence at varying levels of detail over the past 100–150 years.

Source 1 shows how temperatures have changed – and increased – since records began. On average, global land temperatures are 1°C higher now than they were at the end of the nineteenth century. Predictions by a number of leading organisations of a rise of up to 4.5° by the end of the twenty-first century may not seem to be very much, but a recent study of temperatures in the Permian period 250 million years ago suggests that a 6°C rise in temperature led to the extinction of 95 per cent of the species living at that time.

Source 2 shows CO_2 emissions from selected countries. Many scientists believe that it is the increase in greenhouse gases which is the main a cause of global warming. Given the potentially severe consequences of widespread climate change in future, there have been a number of high level meetings to try to decide what action should be taken to cut down greenhouse gas emission, especially CO_2.

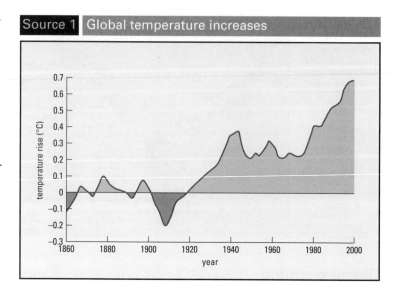

Source 1 Global temperature increases

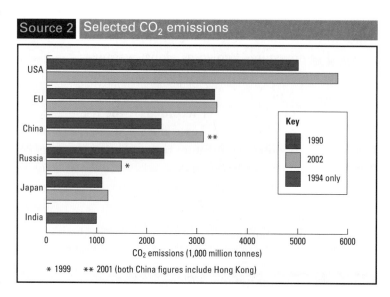

Source 2 Selected CO_2 emissions

Key
1990
2002
1994 only

* 1999   ** 2001 (both China figures include Hong Kong)

Global action

The Montreal Protocol agreed in 1987 is a good example of how international co-operation can reduce greenhouse gas emissions. **Chloro-fluorocarbons** (CFCs) are found in aerosols, coolants and some packaging (Source 3). When released in the atmosphere they can damage the ozone layer which protects us from harmful UV radiation. Gradually CFCs have been replaced by less harmful substances.

World Summits

1992 saw the first world **Earth Summit** on the environment held in Rio de Janeiro in Brazil. It resulted in a range of global policies for future use of resources and the protection of the environment. At the next summit, held in Kyoto, Japan in 1997, a new treaty was set out to try to put into practice recommendations made by the United Nations Framework for Climate Change

Source 3 CFC emissions have been greatly reduced

(UNFCC). This requires countries to cut their emissions of greenhouse gases by an average of 5 per cent by 2012 (based on 1990 figures). The **Kyoto Protocol** was seen as essential to reducing greenhouse gases and mitigating the effects of global warming.

For the treaty to be ratified it had to be signed by at least 55 countries responsible for 55 per cent of 1990 emissions. This has proved to be an impossible target so far, mainly because in 2001 the USA decided it would not sign. Unless virtually all the other industrialised countries sign, the treaty will not become fully operational because the USA is responsible for over a third of greenhouse gas emissions.

Further controversy centres around carbon 'quotas'. Each country has a set quota of carbon emissions which, if they are unlikely to reach, they can trade as 'carbon credits' with countries who are over their targets. Countries can also offset their emissions totals against carbon 'sinks' – their forested areas. This would benefit countries like Russia, very dependent on fossil fuels but with large areas of forest.

Despite initial agreement at Kyoto, the treaty has not yet been ratified and emissions in many countries are still rising, rather than falling. Richer MEDCs do not want to see a reduction in choices and rising energy costs whilst they see that poorer LEDCs have not been set targets – yet it can be argued that global problems can only be solved by taking gobal action.

Source 4 Countries can offset their carbon emissions by planting forests

1 a Describe what we mean by:
    i biodiversity
    ii a biome.
  b What are the two main causes of damage to fragile environments?
  c **i** What is sustainability?
    ii Describe some of the signs of unsustainable use of our planet.

2 a Describe the three main types of soil erosion.
  b **i** What are the main physical causes of soil erosion?
    ii What are the main human causes of soil erosion?
  c What is the difference between a desert and desertification?
  d **i** What are the main physical causes of desertification?
    ii What are the main human causes of desertification?
  e What can be done to help prevent desertification?

3 Study the chart below which shows the rates of deforestation of rainforest in selected areas of the world.

Country	Area of forest lost per year (1000s hectares)
Costa Rica	65
Malaysia	255
Philippines	91
Ivory Coast	290
Nigeria	300
Brazil	1480
Indonesia	600

  a Show the figures on a bar graph.
  b Describe and explain the main causes of deforestation in the Amazon rainforest.
  c Give two reasons why forests need to be managed.
  d Describe how forests can be managed to ensure sustainable development.

4 a What are the four main greenhouse gases and how are they produced?
  b How do greenhouse gases contribute to global warming?
  c What evidence is there for global warming?
  d **i** What may happen in the future if global warming continues?
    ii How might this affect you?

5 a Why is it considered essential to cut down CO_2 emissions?
  b **i** What did the Montreal Protocol help to reduce?
    ii How?
  c Describe the main aims of the Kyoto Treaty
  d Why has the treaty not yet come into force?
  e What are:
    i carbon sinks
    ii carbon quotas?

UNIT 8

Globalisation and interdependence

Unit Contents

The world's major companies (TNCs) have factories in many different countries.
What are the advantages and disadvantages of such globalisation?

The term **globalisation** was first used in the 1960s, but it is since the 1990s that its use has become widespread. One of the problems is that even geographers, economists and others cannot really agree exactly what the term means – whilst in the media the word often seems to be used to describe almost anything which is happening in the world today.

To most geographers, globalisation is the process by which decisions and actions in one part of the world have an important effect on what happens in other parts of the world, regardless of distance or location. As a result of globalisation, many countries have an increasing **interdependence** on others, especially via trade in the global marketplace (Source 1). Increasingly people see the process in the global shift of manufacturing, production and services as a whole range of jobs are outsourced to different countries, for example call centres for USA and UK companies in India and Thailand.

It could be argued that globalisation as a process is not new, as centuries of trading between people and countries could be thought of as globalisation. What has changed since the 1990s is the scale. There are two main reasons for change:

- new technology, which has made physical distances worldwide less important – the 'shrinking world'. From faster and more reliable methods of transport through to better telecommunications and the world wide web,

| Source 1 | Call centres in India may deal with customers for UK companies |

communications of all kinds are cheap, easy and efficient
- the increase of 'free trade' – international trading with few restrictions in terms of tariffs or import duties.

However, whilst many see globalisation as positive, opening up the global marketplace and allowing free flow of goods – and money – across the world, others see it as a way for the richer MEDCs and powerful **transnational corporations** (TNCs) to make even more money at the expense of poorer LEDCs. Anti-globalisation protests have been staged outside the regular meetings of the World Trade Organisation (WTO) (Source 2) since the late 1990s. Many protestors see globalisation as the spread of western culture worldwide, rather than as a purely economic process.

The growth of globalisation has fuelled a major debate about the real benefits of the process. Here are the advantages and disadvantages according to Birdlife International, a conservation organisation whose views are typical of many:

What is good about globalisation?

- People gain access to goods and services from all over the world.
- Greater variety of things to buy and experience from all over the world.
- Benefits can trickle out from investment areas to improve the lives of many, and foreign investment in a country can help a government raise money for healthcare, education, etc.
- The world beyond country borders 'opens up' to us, and we become aware of our role as 'global citizens.'

What is wrong with globalisation?

- Lack of accountability: Multinational companies in place of independent governments become responsible for decisions about where to set up factories, where to close them down, which countries to invest in and how they will treat workers. These multinational companies make these decisions based on what makes the best profit for them, and they may neglect environmental and social justice issues. The government of a country, or a group of workers, may object to some aspect of a multinational company's behaviour. This may then influence the company's decision to move elsewhere (perhaps where people will work for less money), leaving unemployment and chaos behind them.
- The world's poorest have yet to see much benefit from globalisation, and some people argue it makes their position worse. Multinationals tend to locate in already thriving urban areas or close to busy ports, and the wealth generated does not trickle out to the poorer areas. The presence of multinationals can often increase the gap between rich and poor within the same country. The development that comes with globalisation is often not sustainable – any investment can disappear as quickly as it came, if global or local economic conditions change. Free trade only really benefits those who can afford to make, export and buy expensive imported goods, so the poorest are excluded.
- Cultural loss and 'sameness': The presence of global brands and the sophisticated advertising used to sell them makes every street start to look the same – people everywhere (if they can) buy the same burgers and want to wear the same few brands of clothing. Local or national varieties and brands of food, clothing, etc., may not be able to compete, and may be seen by young people in particular as old-fashioned or undesirable. Certain family or religious values may be undermined if the 'foreign' influences become very strong.

Source 2 Anti-globalisation protests outside WTO talks in Seattle, 1999

World trade patterns

Trade is the flow of goods and services between people. There are many different types of trade, as Source 1 shows. International trade involves selling goods to other countries (**exports**) and buying goods from other countries (**imports**). Trade is essential and Source 2 shows why countries need to trade.

Source 2	Why do we trade?

Source 1	Types of trade

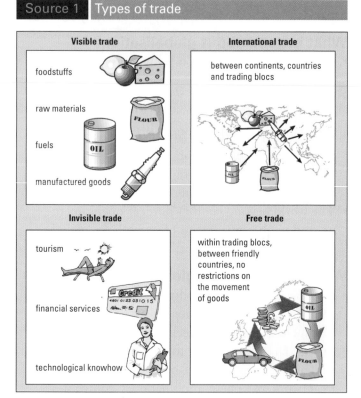

Trade groupings

Some countries have grouped together to form **trading blocs** (Source 3). The European Union is one example. Trading blocs allow member countries to buy and sell goods often with no tariffs (taxes) being charged, so goods traded inside the bloc are cheaper. The countries can sometimes negotiate lower prices for imported goods because the countries act together. The richer countries of the North can often dictate the price they will pay for goods from the LEDCs.

Source 3	Some trading blocs

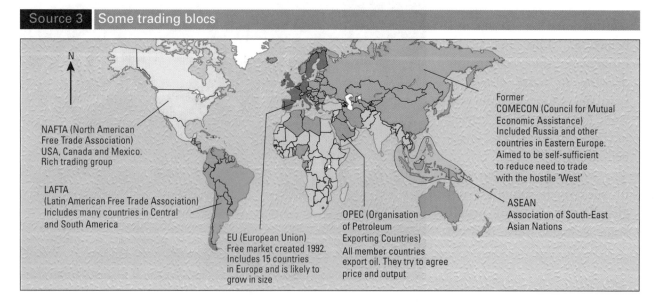

The balance of trade

The **balance of trade** is the difference between the costs of imports and the value of exports. Some countries earn vast profits from their exports and need to import very little. These countries have a trade surplus. They become richer and more developed. Other countries earn much less from their exports, and imports cost much more. These countries have a trade deficit and they become poorer.

The world pattern of trade

The pattern of trade is different for the MEDCs and the LEDCs (Source 3). Most trade is between the richer countries such as Japan, USA and the members of the European Union (EU). The MEDCs have a greater volume of trade and their goods are higher in value. The LEDCs have little to export and their products are relatively low in value as they are usually raw materials or commodities, to which little value has yet been added. Many LEDCs rely on just one or two export products.

The MEDCs rely upon foods, fuels and minerals from the LEDCs. The LEDCs use the money from exports to buy machinery and technology from the MEDCs. This helps the LEDCs to develop. However, the trade is not balanced because:

- the primary products, i.e. fuels, minerals and food, mostly produced by LEDCs, are low in value
- the prices of primary products fluctuate on the world market and are often controlled by the demand from MEDCs
- the value of primary products has not risen at the same rate as the value of manufactured goods
- tariff barriers act against the LEDCs.

In general, MEDCs have a trade surplus. They earn more from their exports than the cost of their imports. However, most LEDCs have a trade deficit. The costs of their imports are greater than what they earn from their exports. This means that many LEDCs have needed to borrow money, usually from the World Bank, to cover the shortfall. Some LEDCs have huge debts and in

| Source 4 | The world pattern of trade |

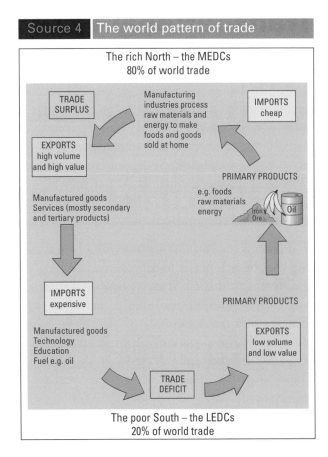

| Source 5 | The trade trap |

recent years many have been unable to repay the interest charges, let alone the borrowed money.

Transnational corporations (TNCs)

One of the biggest changes in industry in the last 50 years has been the way the provision of many goods and services has become increasingly concentrated in the hands of a few large companies. Source 1 names a number of famous companies. What they all have in common is that they are **transnational corporations** (TNCs).

A TNC is a large company which has factories, branch plants or offices in a number of different countries. Many are involved in a range of different economic activities. Unilever is a good example of this type of company. Its headquarters are in Rotterdam in the Netherlands and in London in the UK. These are the places where the most important decisions about the company are made. However, Unilever, which owns many brand names from Bird's Eye fish fingers to Persil washing powder, operates in many countries (Source 2).

Source 1   Some transnational corporations

Source 2   Unilever around the world

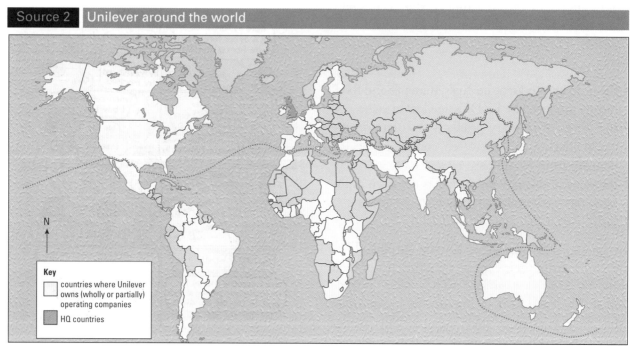

Key

countries where Unilever owns (wholly or partially) operating companies

HQ countries

Many TNCs control the whole production process, from raw materials to the finished product. For example, where a product like a car is assembled or put together in a factory, many of the raw materials will have been gathered together from all over the world.

Source 3	Advantages and disadvantages of TNCs for the host country	
Advantages	**Disadvantages**	
• Develop trade links with other countries	• Often have no regard for the local environment	
• Provide jobs in mines, factories and plantations	• Profits may leave the country	
• Develop infrastructure, like roads and railways	• May use cheap unskilled labour	
• Earn the host country foreign currency when goods are sold abroad	• May close factories and move to somewhere more profitable	
• Bring in professional skills	• May produce goods which are not appropriate to local needs	
• Invest in new technology	• Use of technology may increase unemployment	

As a result, TNCs are extremely powerful. Some TNCs have higher turnovers than some LEDCs. It is not surprising that many poor countries are keen to attract investment from these corporations, but there are advantages and disadvantages which need to be weighed up carefully (Source 3).

Among LEDCs, some countries have attracted more investment than others. Those in the Far East, such as Taiwan, South Korea, Thailand and Malaysia (pages 110–11), are now described as NICs (newly industrialising countries) after much investment by TNCs. For most TNCs, locating in LEDCs cuts costs considerably, especially wages. This global shift in the location began in production and manufacturing, especially goods such as clothes, electronic goods and vehicles. In addition to lower wages, plenty of labour was available and it allowed factories to be built closer to other and new markets, especially South America and Asia. Seventy per cent or more of the major companies in the USA and Europe have outsourced either manufacturing or service jobs overseas. In recent years many service jobs, for example call centres and financial services, have been outsourced from major companies in MEDCS to locations such as India. These use graduate level workers, but afford savings of up to 40 per cent in wage costs.

Company decisions are taken in one country which affect what happens in many other countries. For example, when General Motors took the decision to stop car production in their factory in Luton in early 2001, this decision was made in their headquarters in Detroit in the USA. When someone buys a new car from a multi-national company such as Ford, it is difficult to know where it was made. The car itself could have been assembled in one of a number of different European countries, while the parts will have been made in many different countries.

One indication of the power of TNCs is the way people from many different countries buy and consume the same products. Levis are worn, Ford Mondeos are driven and McDonald's burgers are eaten by people in many different places worldwide.

Source 4	A TNC in Malaysia

The global car industry
Toyota, Japan

Five of the world's top ten transnational corporations are car companies – Ford and General Motors (USA), Toyota (Japan), Volkswagen and Daimler-Benz (Germany). These five alone employ over 1.5 million people worldwide. The car industry was one of the first to locate production away from the company's home country. At first this was just the manufacture of components, but it soon became beneficial to locate new factories overseas. This was to open up the market for their products, avoiding import duties, and to take advantage of lower wages and often substantial financial incentives for bringing work to LEDCs or areas of high unemployment. In the mid-1990s the industry was dominated by the USA, but by the 1990s Japanese companies were competing alongside them.

Toyota

The Toyota group began in 1918 as a spinning and weaving company. Based on the skills and technology gained from manufacturing automatic looms, the Toyota Motor Company was set up in 1937, producing its first passenger car. It helped pioneer the 'just in time' system where parts would be delivered as and when cars were being assembled, so Toyota did not have to store (and pay for) components in advance. This also led to many smaller suppliers locating close to the main factory.

Progress was set back by the Second World War, and Toyota had to almost start again after 1945. By 1947 it had produced its first small car. Rebuilding meant that Toyota's factories were amongst the most advanced in the world. New production methods were introduced including the now famous *kaizen* and total quality control (TQC) systems. The *kaizen* system utilised printed cards attached to each part used, giving accurate stock control information, whilst TQC encouraged any worker to stop production if a fault was found, improving the quality of finished vehicles. Sales grew rapidly, first in Japan and then overseas.

Source 1 Inside a modern car factory in Toyota City

Source 2 Toyota worldwide

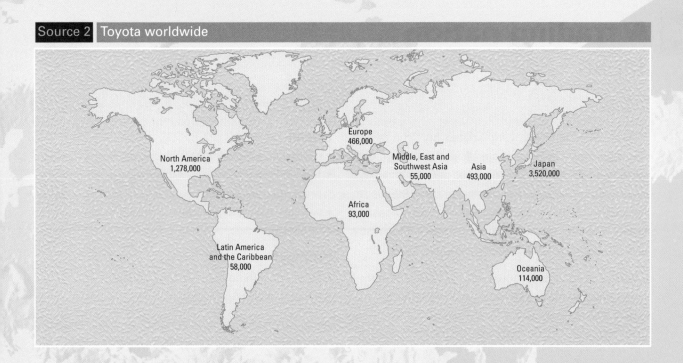

North America
1,278,000

Europe
466,000

Middle, East and
Southwest Asia
55,000

Asia
493,000

Japan
3,520,000

Africa
93,000

Latin America
and the Caribbean
58,000

Oceania
114,000

Fact File Toyota

Main brands: Toyota, Lexus, Daihatsu and Hino
2003 production: 6.68 million vehicles
Japan: 12 car plants, mainly around
                          Toyota City (near Nagoya)
Overseas: 51 manufacturing companies
                          in 26 countries
Employees: 264 000

Source 3 Toyota overseas manufacturers: country and number of factories (planned developments in brackets)

Country	Factories	Country	Factories
Argentina	1	Mexico	0 (1)
Australia	1	Pakistan	1
Bangladesh	1	Philippines	2
Brazil	1	Poland	1 (1)
Canada	2	Portugal	1
China	9 (3)	South Africa	1
Colombia	1	Taiwan	2
Czech Republic	0 (0)	Thailand	2
France	1	Turkey	1
India	2	UK	1
Indonesia	2	USA	7 (1)
Kenya	1	Venezuela	1
Malaysia	1	Vietnam	1

From the 1970s Toyota started to concentrate its operations overseas, for the reasons given in the opening paragraph. By 1991 it had become (and remains today) the third largest motor vehicle manufacturer in the world and had overtaken General Motors to become the world's largest producer. The mid-1990s saw a decline in global car sales. Toyota continued to invest abroad (Source 2) where it has fostered partnerships with local companies to supply components and assemble finished vehicles.

Toyota in Pakistan

Toyota began production in Pakistan in May 1993. It has a quarter share in the Indus Motor Company (IMC). Half the company is owned by the Pakistan group, House of Habib, and chaired by Ali S. Habib. The factory covers a 105-acre site in the Port Bin Qasim Industrial Zone outside Karachi. The factory produces six models in the Toyota Corolla, Toyota Hilux and Daihatsu Cuore range, employing 1100 workers in a modern factory, using the Toyota Production System. It is the only distributor of Toyota cars in Pakistan. When Toyota celebrated ten years of production in 2003, the factory had built just over 100 000 vehicles. Having taken four years to produce the first 25 000 cars, in the next three years it produced 50 000. In 2003 it produced 20 000 cars.

A trading nation
Japan

Before 1945 Japan was very isolated from the rest of the world. Few foreigners were allowed entry and there was little trade. Since 1945 there has been an economic miracle. The GNP has risen greatly and in 1989 Japan replaced the USA as the world's richest nation despite having very few natural resources. Trade and industry have developed for the following reasons:

- Japan has no oil or iron ore, very little coal or other raw materials, and needs to import these items
- Japan has to export to pay for the imports. Steel, chemicals, cars and ships were sold abroad. Since the 1970s these industries have declined. They have been replaced by a growth in the electronics industry and in services
- The Japanese have developed and manufactured many new products, for example computers, video recorders, compact disc players and video cameras. Source 1 shows the main imports and exports of Japan today.

The reasons for Japan's economic miracle are as follows.

Economic
- modern machines and methods of working
- profits used in research to develop new products
- large home market which has become richer

Social
- well-educated workforce
- the workforce operates well in teams and are prepared to work long hours
- workers have a high degree of loyalty to their employers

Political
- strong government support for industry
- political stability

Japan has been so successful that every year there is a huge trade surplus. It was $135 billion in 1995. The country is very dependent upon the rest of the world to supply the raw materials that its industry needs, and to provide the market for the goods it makes. Japan is **interdependent** with many countries.

Japan's success has also brought some problems. Japan has been in trouble in the world over its trading policies. Japan has the benefit of free trade in many areas of the world yet charges high tariffs on goods entering Japan.

Source 1 Main imports and exports of Japan

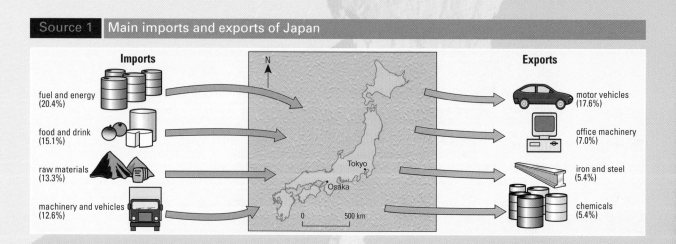

Global tourism 8.6

Tourism is the world's fastest growing industry, and in many countries, especially LEDCs, is the biggest employer and biggest contributor to GDP. Source 1 shows just how rapidly tourism has grown, from just 20 million arrivals in 1950 to over 700 million in 2003. The graph, produced by the World Tourism Organisation also shows the predictions for future growth with an estimated 1.6 billion tourist movements by 2020. This is a growth rate of just over 4 per cent per year between 1995 and 2020.

There are many reasons why tourism numbers have increased so rapidly, especially in the last 50 years:

- Travel, especially by air, is easier, cheaper and quicker.
- Many people have longer, paid holidays.
- Many people have more money to spend and more leisure time.
- Package holidays have attracted many people to travel abroad.
- Countries see tourism as a valuable source of income and have built up their own tourist industry.
- Many people are taking early retirement and/or live longer and want to travel.
- There are many specialist holidays covering a wide range of interest.
- Even the world's remotest places are now accessible, for example the Antarctic, Amazon rainforest.

Although there have been economic benefits from increased tourism, it has also put great pressure on the areas receiving visitors. This is causing particular difficulties in areas which attract high numbers of visitors (mass tourism), for example Spain's Mediterranean coast, and in environmentally sensitive areas, once remote and very difficult to access, such as Machu Picchu in Peru. This has helped encourage the growth of **ecotourism** where tourists visit locations in small

| Source 1 | World tourism arrivals by region, 1950–2020 |

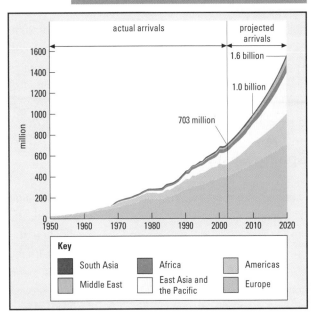

| Source 2 | Machu Picchu on Peru's Inca Trail is suffering from too many visitors |

groups. Money goes to the local communities rather than to hotel owners or large companies, and benefits local people in addition to providing jobs. Other areas under pressure are sites close to large areas of population, for example the National Parks in the UK.

Mass tourism
the Mediterranean coast

Europe is by far the world's most popular tourist destination, with four countries in the top ten for 2000 (Source 1) and about 60 per centof the global market. The majority of these tourists visit either the mountains or coastal areas. The Mediterranean attracted 135 million tourists in 1990, rising to 220 million in 2002. It is estimated that this will rise to 350 million by 2020.

Spain's Costa del Sol

Spain's Costa del Sol is a 160 km stretch of Mediterranean coast (Source 2) stretching either side of Malaga. In the 1950s it was a fairly quiet coastal area, relying heavily on fishing. However, since the growth of cheap package holidays in the 1960s (many arriving via flights into Malaga), the area has been completely transformed. Today it is an almost unbroken strip of high-rise hotels, holiday apartments, shops, cafes and restaurants. Millions of tourists are attracted by the hot, dry, sunny weather and a wide range of leisure facilities including water sports and over 60 golf courses.

| Source 1 | The top ten tourist destinations, 2000 |

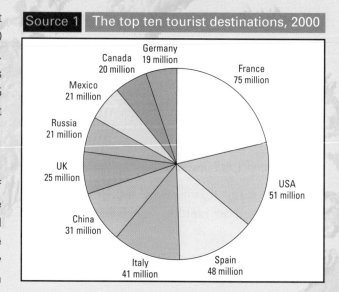

Germany 19 million
Canada 20 million
France 75 million
Mexico 21 million
Russia 21 million
USA 51 million
UK 25 million
China 31 million
Spain 48 million
Italy 41 million

This growth has brought many advantages to this part of Spain, especially jobs in the tourist industry itself. More recently the infrastructure of the region has seen a number of improvements, benefiting visitors and local residents alike. This is part of the multiplier effect – more tourists arrive, spending more money which

| Source 2 | Spain and the Costa del Sol |

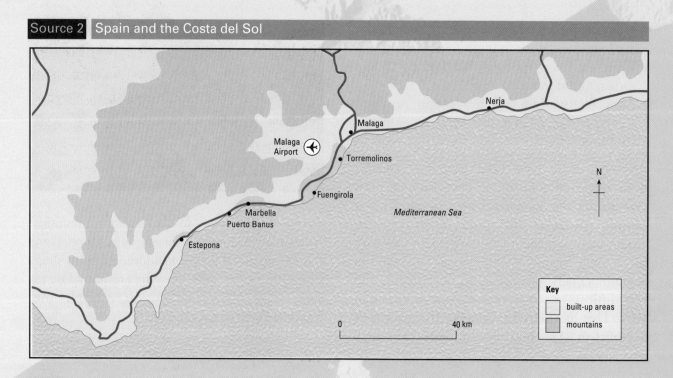

Nerja
Malaga
Malaga Airport
Torremolinos
Fuengirola
Marbella
Puerto Banus
Estepona
Mediterranean Sea
N

0   40 km

Key
built-up areas
mountains

creates jobs, not just in tourism but in the construction industry and for local suppliers, for example farmers.

However, such high numbers of visitors concentrated mainly from May to October, visiting a relatively small area around the Mediterranean coast, has increased pressure on what are often quite limited resources. This includes:

- high demand for water in areas where it is often a scarce resource. Tourists typically use almost twice as much water per day as local residents.
- the production (and disposal) of over 40 million tons of waste each year
- increased urbanisation of coastal regions as more hotels and tourist facilities are built, damaging local ecosystems
- the increase in the number of second or holiday homes, which take up much more land than hotels but are usually only occupied for short periods
- high levels of pollution, particularly from cars, aircraft and boats.

Protecting the coast

For many years during rapid developments in the industry, there has been no overall strategy for the development of tourism along areas like the Costa del Sol. In 1975 the Mediterranean Action Plan (MAP) was set up, initially to protect the marine environment. In 1995, MAP widened its brief to cover the sustainable development, biodiversity and management of coastal regions.

Source 3 | Tourists enjoy the warm Mediterranean climate

Some areas, such as the Balearic Islands off Spain's Mediterranean coast, are looking to introduce an eco-tax paid by all visitors to help fund environment-friendly developments. This could help many tourist areas in the Mediterranean, especially if the money raised is reinvested in protecting and managing the some of the more environmentally sensitive coastal areas. Centres like Benidorm have invested millions in improving sanitation and water supply, and the once notorious N340 highway has been made considerably safer by rebuilding and road widening. The Spanish government has also brought in much stricter planning and building regulations, and demolished some sub-standard developments.

Source 4 | High rise hotel developments along the Costa del Sol

Sustainable tourism in a LEDC
Tanzania

Source 1 Coastal Tanzania

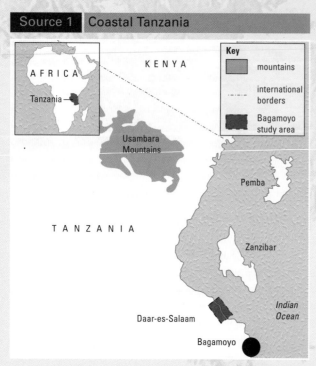

Tanzania is one of the poorest countries in Africa. It has a Gross Domestic Product per person of only $700. Tourism is one way to develop the country. If more foreign tourists come to Tanzania they bring much needed dollars and other foreign currency to spend.

Tanzania has a coastline and islands with sandy beaches and coral reefs which give tourists the chance to relax and explore. If coastal tourism is to make a large contribution to development it must be managed. Sustainable development will depend on preserving the coastal resources.

Bagamoyo – can tourism take the pressure off coastal resources?

South of Dar-es-Salaam the coastal area of Bagamoyo is under siege. Mangroves, salt and coral reefs provide the large and growing population with their main resources outside peasant farming.

Mangroves provide wood for building as well as fuel wood, charcoal and fishing stakes. Medicinal plants are also collected. Too much mangrove forest is being removed. Areas of mangrove swamp are left to

evaporate to produce salt. Now these salt ponds are in decline as cheaper imported salt has become available.

The coral reefs around Mwamba Kuni and Mwamba Mshingwi, and elsewhere, are also under severe threat. The local people depend upon fishing. The waters around the reefs are the main fishing grounds. Illegal dynamiting, trampling on the coral and removal of shells for sale are destroying this resource. At the same time the use of fine mesh nets means that juvenile fish are caught before they have time to reach maturity and breed.

Source 2 The location and resources of the Bagamoyo area

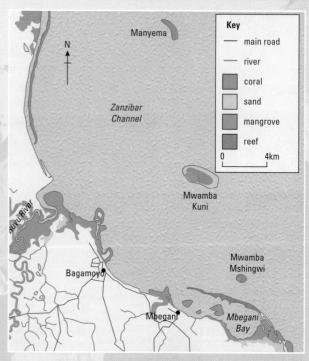

At present only five hotels have been built. The local fishermen are prevented from using the waters in front of the hotels. Little employment is to be found in the tourist industry as yet. However, as the resources dwindle it may be that a more developed tourist industry will not only provide more jobs, but also help protect these endangered resources.

Zanzibar – the tourist paradise?

The recent growth in foreign tourists has been dramatic – an increase of over 300 per cent. Foreign currency earnings have swelled from US $13 million in 1970 to US $730 million in 1999. The island of Zanzibar is a fast-growing tourist 'honeypot'. It needs to be. The population is growing at 3 per cent per annum and by the year 2015 is expected to reach 1 500 000. This will mean a population density of over 600 per square kilometre.

| Source 3 | Growth of tourism in Tanzania 1970–1999 |

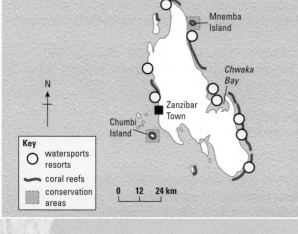

Pemba – the future paradise?

North of Zanzibar is the relatively undeveloped island of Pemba. Like Zanzibar, it is blessed with excellent coral reefs. Unlike Zanzibar, it is not yet a tourist 'honeypot'. There are few hotels on Pemba, but the first high-quality hotel has just been opened.

Pemba has the opportunity to limit the pressures on the environment and encourage more environmentally friendly tourism. Masali Island off the west coast is a coral-fringed, protected nature reserve. The newest hotel, Fundu Lagoon, is built to blend in with the forest and mangrove trees – a good example of **ecotourism** (Source 7).

Pressure on the beaches and coral reefs is growing. Most resorts offer diving and snorkelling. The reefs must be protected if tourism is to continue. Some 23 000 local fishermen rely on the coral reefs as their fishing grounds and provide the main source of protein for local people. Both the tourists and the fishermen are threatening the reefs. Coral is damaged by tourists who trample on it and collect it illegally. The reefs are damaged by boat anchors and polluted by hotel effluents. Now the government is creating protected areas to safeguard these fragile resources.

| Source 4 | Water-based tourist resources on Zanzibar |

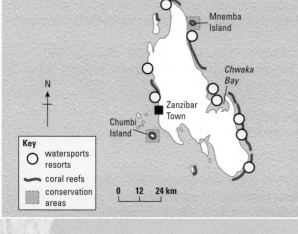

| Source 5 | Zanzibar offers tropical beaches and coral reefs |

| Source 6 | Water-based tourist resources on Pemba |

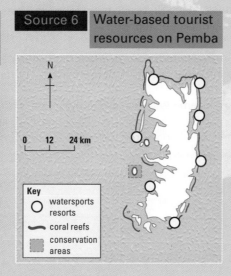

| Source 7 | Blending tourism and environment – ecotourism on Pemba |

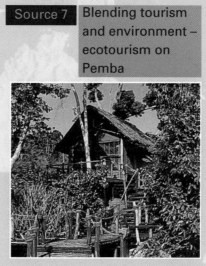

1 a What is meant by:
  i globalisation
  ii interdependence
  iii a TNC?
b Name two of the main reasons for the growth and development of globalisation.
c Define the following terms: trade; international trade; exports; imports; balance of trade; trade deficit; trade surplus; trading bloc.

2 a State two characteristics of transnational companies.
b Make a table like the one below and fill it in for five transnational companies.

Name of transnational company	What it makes or does

c Many transnational companies have set up factories in LEDCs. State three advantages of this for LEDCs.
d Transnational companies are said to operate globally. Give as many examples as you can to show how they operate in a global way.

3 a What were the main reasons for the world's major car manufacturers to locate their factories overseas?
b The Japanese car company Toyota run their factories using the Toyota Production System. Explain what this is and how it works.
c Draw a bar graph showing car production in each continent for Toyota in 2003.
d What are the advantages and disadvantages for Pakistan from the location of the Toyota car plant in Karachi?

4 a Why is tourism such an important global industry?
b Why have tourist numbers risen so quickly?
c What is ecotourism?

5 a Describe why tourism is so important to the economy of Tanzania.
b In what ways are the coastal resources under threat in Tanzania?
c Define what you understand by the term 'sustainable development' and what Tanzania must do to sustain its tourist industry.

Human welfare

Unit Contents

These people receive few of the world's resources. How good are their lives?

Development and human welfare

Countries and regions develop, including economically as Unit 4 shows. People have a quality of life and a level of happiness which we call human welfare. Development does not necessarily benefit ordinary people's happiness and quality of life. Development studies rely heavily on GDP and what the money will buy, for example doctors, cars, school places, etc. Human welfare is an extension of development into looking at a fuller picture of how well people live in different places. As we all know, money is not everything in life as recent research shows - the quality of life in the UK was higher in 1976 than it is now despite a 50 per cent rise in real GDP since 1976. Happiness and people's satisfaction with their lives does not necessarily rise with increasing prosperity and more goods and services.

It is usual in human welfare studies to use various indicators of development and quality of life and roll them into an index like the Physical Quality of Life Index (PQLI). These indexes generally combine not only economic indicators, such as GDP, with population and social indicators, such as housing, health and access to basic services like electricity and clean water, but also with environmental indicators. Pollution levels, wild bird populations and the chance of being victims of crime also affect people's welfare.

Measuring quality of life

The Physical Quality of Life Index (PQLI) was introduced in 1979, using three indicators (Source 1), put together on a 0–100 scale to give a better indication of well being. They are all social measures, predominantly health-based, so PQLI should give a more accurate impression of human welfare than just GDP.

Source 1	Measuring quality of life
PQLI	Basic literacy Infant mortality Life expectancy and aged 1 year
HDI	Life expectancy at birth GDP Adult literacy and school enrolment
HPI	Probability at birth of not reaching 40 years Adult literacy % without access to clean water % of underweight children under 5 years old

In the 1990s the UN introduced the Human Development Index (HDI) (see Unit 4, page 98) which also used a combination of social and economic indicators. It also introduced another formula, the Human Poverty Index (HPI), this time using indicators for longevity, adult literacy, access to water and underweight children.

Today, HDI is probably the most commonly used general method of comparing the quality of life and development level of different countries. However, even the UN concedes that neither HDI or HPI include other important factors which contribute to people's overall quality of life. Many of these are difficult, if not impossible, to measure in terms of hard data, for example political freedom, security and the freedom to follow religious or cultural practices.

A range of economic, population and social indicators for selected countries can be found in Unit 4.1 Development indicators, on pages 98–101.

Variations in human welfare

There are great variations in human welfare and quality of life across the world, both between and within countries. Clearly, one cause is the different levels of development that exist between areas and countries.

Economic development does provide the opportunity to improve the quality of people's lives and can lift people out of poverty. It provides the resources to invest, for example, in girls' education and healthcare for babies, which do benefit the lives of ordinary people. However, governments with inappropriate spending priorities could see the money being spent instead on things such as military jets and helicopters which do little for quality of life. Generally, GDP and quality of life have a strong, though not perfect, relationship. Richer countries tend to have a higher quality of life though not necessarily happier people than poorer countries.

Bad government is a major human cause of poor levels of human welfare and quality of life. Nelson Mandela, the former President of South Africa, recently said that poverty is people-made. Incompetent or corrupt governments lower the welfare of their populations by:

- spending money on themselves
- putting aid money into large, prestige projects
- allowing legal systems in which the public have no confidence to continue
- failing to provide an adequate and reliable infrastructure of electricity, clean water and roads
- ignoring pollution
- pursuing conflicts and war, and turn off foreign investors.

The absence of free and fair international trade helps to keep quality of life low in LEDCs. At present many LEDCs cannot compete with richer MEDCs who frequently subsidise their own industries or dump surplus produce – usually foodstuffs – on the world market at low prices. Both the Fair Trade movement and Oxfam have taken a lead in this. Free trade is the ability to buy and sell goods freely across the world, without duties, tariffs or import restrictions. The Fair Trade movement deals directly with farmers in LEDCs, paying a fair price for their products rather than them having to sell through world commodity markets or to large TNCs (Source 1).

The freeing up of international trade to give LEDCs greater access to MEDC markets would be a big step on the way to improving quality of life in many LEDCs.

Other causes of welfare differences can be due to the physical environment. Scandinavia's cold climate, for example, means more resources have to be put into heating, clothing and housing, leaving fewer for other aspects of quality of life. The lack of easily available water supply in countries around the Persian Gulf, the threat of hurricanes in Caribbean islands and the damage caused by earthquakes and volcanic eruptions in countries on plate margins, lowers quality of life there.

Aid is the transfer of money, goods and expertise to assist the development of LEDCs and improve the quality of life in these countries (Source 1).

Source 1 | Types of aid

Official Development Aid (DFID)
This includes bilateral grants, loans and technical assistance. This is the aid which goes from government to government. The LEDC designs the development scheme.

Voluntary aid
This is given by individuals rather than governments to national and international charities, for example Oxfam, Red Cross.

Multilateral aid
A country provides aid through a third party such as the United Nations, the World Bank or the IMF (International Monetary Fund). This is a growing source of aid and is not tied to any one country.

Emergency short-term aid
This is aid in the form of food, shelter, medical supplies and water that is sent to countries following a natural disaster.

Aid can come from two main sources:

1 Governments Governments are involved in two types of aid. The first is **bilateral aid** where the government of the country gives the aid directly to the government of the receiving country. The aid may include grants, loans and technical assistance.

The second main source is **multilateral aid** where a government donates aid to large international organisations such as the World Bank or the United Nations. These organisations then send the aid to the countries in need.

Some governments give aid with strings attached. This is called tied or conditional aid because the donor countries look for something in return, perhaps a military base, the purchase of weapons or a trade agreement. For example, the USA gave Peru large amounts of aid to search for oil. In return Peru purchased jet aircraft from the USA and allowed fishing boats from the USA into their waters.

2 Non-governmental organisations (NGOs)
NGOs are charitable organisations such as Oxfam, the Red Cross and Save the Children. They rely on donations and fund-raising events to pay for projects in poorer countries. It is sometimes called **voluntary aid** because the money is given by individuals. Projects tend to be on a smaller scale but have more of a direct effect on the lives of the local people in the LEDCs.

As well as long-term development projects, short-term **emergency aid** (food, shelter, medical supplies, etc.) is given following a natural disaster. Mozambique, for example, received emergency aid during the floods in 2000.

Source 2 | An example of appropriate aid using intermediate technology

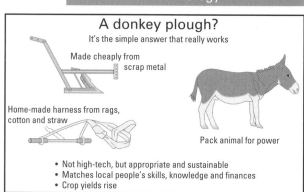

A donkey plough?
It's the simple answer that really works

Made cheaply from scrap metal

Home-made harness from rags, cotton and straw

Pack animal for power

• Not high-tech, but appropriate and sustainable
• Matches local people's skills, knowledge and finances
• Crop yields rise

The benefits and problems of aid 9.4

Giving aid has both advantages and disadvantages for the receiving countries.

Benefits to the receiving country

Well-planned medium- and long-term aid programmes help a country to develop and to become less dependent on aid. For example, schemes in Ethiopia are helping the people become less dependent on emergency aid when a drought occurs. The schemes also:

- improve the education and skills of the people
- increase crop yields to feed the local population rather than growing crops for export
- encourage small-scale industries to be set up
- improve water supplies and health care.

When a natural disaster strikes such as a flood or hurricane the short-term emergency aid, including food, shelters, medical equipment and clothing, is often essential in saving lives.

Problems for the receiving country

Poorly-planned aid often makes a country more, not less, dependent on others. The large prestigious schemes such as the Itaipu HEP scheme in Brazil or the Aswan High Dam in Egypt often have more disadvantages than advantages for the local people who:

- lose their farm land and have to buy food
- need expensive pumps and irrigation systems to access the water
- have to purchase fertilisers instead of being able to rely on natural flooding
- need to purchase expensive equipment, fuel and spare parts from MEDCs.

Small-scale technology that used local resources and was in keeping with the skills, education and financial resources of the local people would have been more appropriate. Appropriate aid is when the technical expertise and equipment given properly suits the conditions in the receiving country (Source 2, page 198).

Look at the schemes in Sources 1 and 2. Which scheme do you think benefited the local people most?

Source 1 | A small-scale well system

Source 2 | Aswan High Dam

Some countries give tied aid and expect something in return for the aid that is given such as trade agreements or military bases. Sometimes the aid is not free and has to be paid back as a loan, pushing the LEDCs into more debt.

Some aid does not reach the poorest people in a country due to poor transport arrangements or corrupt governments who may spend the money on the armed forces, on cities or on tourism. As a result the rich get richer and the poor get poorer.

If MEDCs suffer an economic recession, less money is given to aid projects. Countries are interdependent, each depending on the other.

A UN aid agency
the World Bank

The World Bank was formed in 1944, one of the specialised organisations set up by the United Nations. It is not really a bank. The 184 member countries are responsible for how money is raised and spent, but the main aim of the World Bank is to help reduce poverty. It has specific targets for education, infant mortality, maternal health, disease and access to water, set out in the Millennium Development Goals to be reached by 2015 (Source 1).

The World Bank Group is made up of five different agencies, each with a different purpose. The International Bank for Reconstruction and Development (IBRD) and International Development Association (IDA) are referred to as the World Bank, responsible for loans and borrowing. The International Finance Corporation (IFC) raises private sector funding; the Multilateral Investment Guarantee Agency (MIGA) guarantees investors against loss and the International Centre for Settlement of Investment Disputes (ICSID) arbitrates over investment disputes when necessary.

The Bank has a governing body which meets every autumn, with each member country represented. A smaller group of 24 directors meet twice a week to deal with the Bank's main business, headed by a President who is elected for a five-year term. Each year they lend between US$15–20 billion for a wide range of projects in over 100 countries, the largest provider of development assistance in the world (Source 2).

The Bank has not been without critics, and has listened to public criticism about the pressure of servicing debt in many LEDCs. It has invested heavily in the Heavily Indebted Poor Countries Initiative (HIPC), helping 26 countries by cancelling debt payments provided that money is used for welfare programmes, housing, education and health for the poor, for example in Uganda. It has also taken a lead role in funding HIV/AIDS programmes, especially in Sub-Saharan Africa, helping Nicaragua rebuild after Hurricane Mitch and funding education for girls in Bangladesh.

Source 1	The World Bank's current 'top ten' functions

The World Bank …

1. is the largest external funder of education
2. is the largest external funder of the HIV/AIDS programme
3. is a leader in the aniti-corruption effort
4. strongly supports debt relief
5. is one of the largest funders of biodiversity projects
6. works with partners
7. helps bring clean water, electricity and transport to the poor
8. involves civil society in every aspect of its work
9. helps countries emerging from conflict
10. is responsible to the voices of poor people

Source 2	A World Bank coffee project, Rwanda

Non-governmental organisation (NGO)

Oxfam

Oxfam is probably the UK's – and one of the world's – best known charities. It was set up in 1942 as the Oxford Committee for Famine Relief to help provide relief for the people of Greece under Nazi occupation. Today it operates in over 80 countries, employing 3500 full time staff and thousands of volunteers, many of whom help to run over 800 charity shops in the UK. It is a non-religious NGO working with a range of partner organisations.

Oxfam's main purpose is to work with others 'to overcome poverty and suffering'. Its stated aims are to 'ensure that every individual is assured of:

- a sustainable livelihood
- education and health
- life and security
- a right to be heard
- a right to equity.'

Oxfam is an independent charity, part of Oxfam International and works as part of a worldwide movement to try to build a just and safer world. Its funding comes from donations and it is accountable to its supporters and the people it helps. Whilst Oxfam often offers emergency support in response to disasters across the world, for example sending blankets and medicines in the aftermath of earthquakes, it also works 'in the field' to help improve the lives of some of the world's poorest people. This is often to help establish local water supplies, digging wells and boreholes to provide access.

Oxfam regularly runs specific campaigns. Some of its current campaigns include:

- *Cut the Cost* – to cut the price of medicines in poorer countries, increasing access
- *Conflict* – to reduce homelessness caused by war and armed conflict
- *Education Now* – putting pressure on world leaders to make education available for all children
- *Make Trade Fair* – campaigning for changes to trade rules which favour TNCs and MEDCs at the expense of LEDCs.

Source 1 Where the money comes from

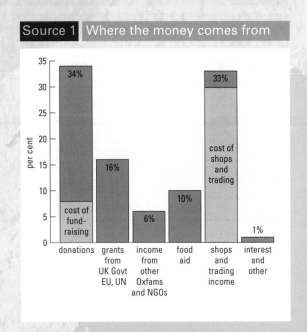

Source 2 How Oxfam spends its money

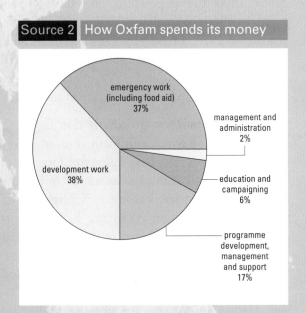

World population growth

Source 1 World population growth 1950–2075

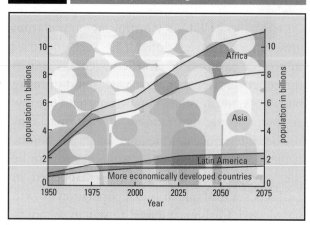

Source 2 Birth and death rates in Burkina Faso, 1950–2000

Year	Birth rate per 1000	Death rate per 1000	Natural increase per 1000
1960	51	32	+19
1970	51	29	+22
1980	50	24	+26
1990	47	20	+27
2000	47	17	+30

Population growth is one of the most important issues facing the world this century. This is shown in Source 1, with the population set to increase from 6 billion to 10 billion in the next 50 years. Since 1800 the population growth has been rapid. In 1960 the total was 3 billion; 4 billion was reached in 1975; 5 billion by 1987 and 6 billion by 1999. This means that the total doubled in just 39 years, between 1960 and 1999. Nearly half the world's population is now under 25 years of age.

This rapid population growth is due largely to many LEDCs having high birth rates at a time when their death rates have fallen to a lower level. Source 2 shows that this has happened in Burkina Faso in West Africa, leading to a large natural increase in population. Why are birth rates so high in these countries? Part of the answer is that people see children as a financial asset. They can help with the work, especially on family farmland, and later look after their parents when they are old. Medical advances, improvements in diet and developments in public hygiene and sanitation, especially clean water supply, explain the drop in the death rate.

Sooner or later the planet will reach its **carrying capacity**. This is the total number of people that can be supported by natural resources. This is not just the land needed to live on, but the amount needed to provide food, water and other resources we find essential to everyday living. The problem of resources is exacerbated by their disproportionate use. MEDCs have a minority of the population, but use the majority of the world's resources. For example, if consumption continues at the present rate, the USA will use more oil than the entire population of Sub-Saharan Africa.

It is important that a country does not exceed its carrying capacity. A growing population can lead to an imbalance between population and resources. Finding and keeping a balance, especially between population and food supply, is vital. Looking at Source 4 you can see that overpopulation occurs when there are more people than resources. Overpopulation puts pressure on:

- the provision of housing
- the provision of jobs
- social services, for example childcare, education
- the infrastructure, for example transport, electricity, water.

Homelessness, squatting, poor quality and temporary housing, unemployment, malnutrition, famine, poverty and low quality of life are likely in overpopulated areas, especially in LEDCs. Having an optimum or ideal size of population for an area's resources offers a sustainable future. Some

Source 3 It is important that girls are well educated and encouraged to stay in school

Technology helps in achieving an optimum population by providing ways of slowing population growth and/or increasing the supply of resources, especially food.

Poverty contributes to continuing population growth in LEDCs. Continuing population growth makes it very difficult for such countries to provide clean water, electricity, schools and all the other services that their young and growing population need. Young populations tend to grow as young people start families of their own. This has been described as a 'demographic time bomb'. Educating and empowering women seems to be the way forward. Where this is happened in LEDCs, for example in India, population growth has slowed. Girls encouraged to stay in school to upper secondary level are:

- more likely to choose contraception and family planning
- less likely to have children when in their teens
- more likely to want to have a career (Source 3).

Source 4 Population and resources

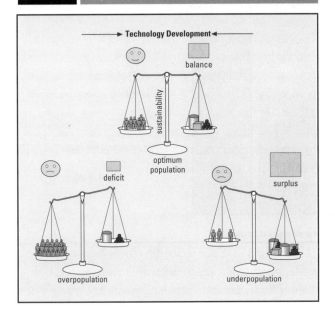

Source 5 Environmental damage from rapid population increase

resources can be used for present economic gain and some conserved for the future. Growing populations need more food. Providing more food can further damage the environment (Source 5).

Population growth
China

China's population in 2003 was 1.3 billion out of a world total of 6 billion, making it the world's most populous country. For years Chinese governments saw their rapidly increasing population as a major asset. After all, a country with between a quarter and a fifth of the world's population cannot be ignored by the rest of the world, even if by world standards the country is relatively poor.

By the 1970s the Chinese government had to do something to try to take control of rapid population growth. Falling death rates, the result of better medical care, and rising birth rates were producing a high rate of natural increase. This was putting a great strain on basic resources like food and water in a country which regularly suffered from both drought and flooding. At this time, the average family size was 3. Source 1 shows what the Chinese government feared would be the effect by 1990. Much of the north of China is arid, and despite great efforts by farmers to produce more food, soil erosion and pollution from fertilisers and pesticides had started to take their toll on the fertility of the land. As China's industries grew, so did the competition between factories and farmers for often limited water resources and the level of pollution, mainly from burning fossil fuels.

Source 2 shows the actions taken, starting with a nationwide family planning programme. However, it was the introduction of the 'one child' policy in 1979 that caused most controversy. Couples were only allowed one child. If they followed this policy, they were entitled to free education, welfare support and better housing. If they had more than one child, they would be fined and have benefits withdrawn. If women were found to be pregnant with a second child, they were often coerced into having an abortion. Many others were sterilised. This ruthless policy resulted in widespread abortions if women found out that they were carrying a female child. Others gave birth and abandoned their female babies. This was because a son is valued more highly – families wanted a boy to work on the farm or to look after parents in their old age.

Fact File	China's population

- 1 in 5 of the world's population live in China.
- 117 boys are being born to every 100 girls.
- Almost 52 per cent of the population are male.
- Most provinces now have a 110:100 ratio of male:female babies or higher.
- By 2020, there will be 40 million batchelors as there are not enough women.
- 500 000 children are abandoned each year, 95 per cent of them girls
- Eighty per cent of China's people live in rural areas, many in poverty.

Source 1	Predicted effects of continuing Chinese population growth

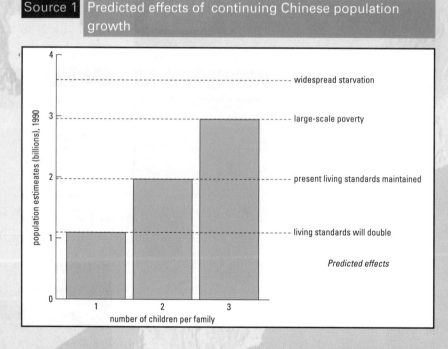

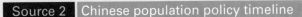

Source 2   Chinese population policy timeline

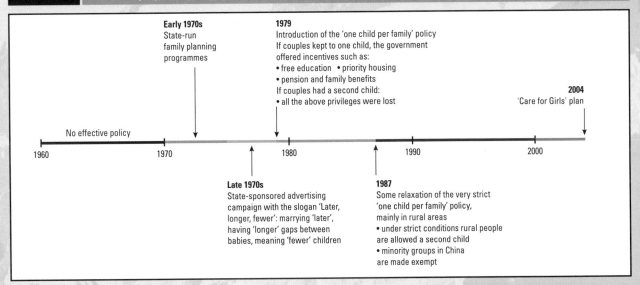

Early 1970s
State-run
family planning
programmes

1979
Introduction of the 'one child per family' policy
If couples kept to one child, the government
offered incentives such as:
• free education • priority housing
• pension and family benefits
If couples had a second child:
• all the above privileges were lost

2004
'Care for Girls' plan

No effective policy

1960   1970   1980   1990   2000

Late 1970s
State-sponsored advertising
campaign with the slogan 'Later,
longer, fewer': marrying 'later',
having 'longer' gaps between
babies, meaning 'fewer' children

1987
Some relaxation of the very strict
'one child per family' policy,
mainly in rural areas
• under strict conditions rural people
are allowed a second child
• minority groups in China
are made exempt

The first softening of the one child policy took place in 2001 when families in rural areas were allowed to have a second child if the first was a boy. Ethnic minorities were also allowed two or three children. They make up only a very small percentage of China's population.

Further relaxation of population controls were announced in 2004 with the announcement of the 'Care for Girls' plan. Worried by the increasing imbalance between male and females (Source 3), which had reached a ratio of 140:100 in some provinces, this aims to 'create a favourable environment for females'. Families with girls will now get free schooling for them and better housing and employment. In one province, Fujian, US$24 million has been set aside for the half a million families with daughters.

As China become more industrialised and certain sectors of the population become more affluent, the pressure on resources will increase. However, the government is beginning to accept that the way ahead appears not to be a one child policy but better education, health care and family planning.

The 'economic miracle' now underway in China may not have been possible without the earlier control of its exploding population. By 2004, the Chinese population was 1.3 billion, roughly 10 per cent more people that the government in 1975 wanted for living standards to be double by 1990 (Source 1). Years of rapid economic growth have seen living standards and the quality of life more than double since 1975, especially in urban China. Poverty has declined throughout China.

Source 3   China's population pyramid

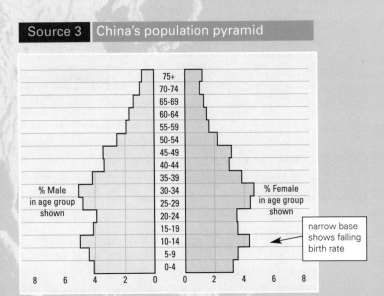

narrow base
shows falling
birth rate

Disparities in human welfare 1
Brazil

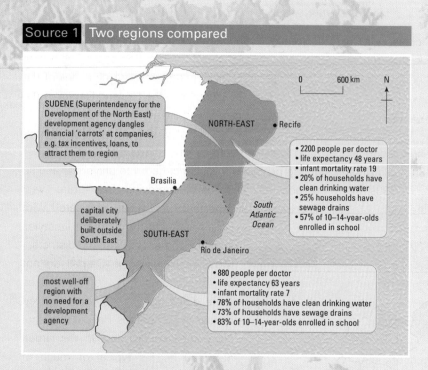

SUDENE (Superintendency for the Development of the North East) development agency dangles financial 'carrots' at companies, e.g. tax incentives, loans, to attract them to region

NORTH-EAST • Recife

0   600 km   N

• 2200 people per doctor
• life expectancy 48 years
• infant mortality rate 19
• 20% of households have clean drinking water
• 25% households have sewage drains
• 57% of 10–14-year-olds enrolled in school

Brasilia

capital city deliberately built outside South East

South Atlantic Ocean

SOUTH-EAST

Rio de Janeiro

most well-off region with no need for a development agency

• 880 people per doctor
• life expectancy 63 years
• infant mortality rate 7
• 78% of households have clean drinking water
• 73% of households have sewage drains
• 83% of 10–14-year-olds enrolled in school

The various figures show just how different quality of life is in these two regions. The North East has double the birth rate and infant mortality compared to the South East. It also has a much lower life expectancy and literacy rate. Half of its population live in rural areas, mostly surviving as subsistence farmers. Access to basic services such as water and healthcare is also much lower. A fifth of all families suffer from hunger and malnutrition on a regular basis.

As a result of these differences, large numbers have migrated from the North East to the South East, especially to the cities (see Unit 6.4, pages 142–3) where employment opportunities are greater. Although this has created problems in the South East, such as lack of adequate housing, water and sanitation and more congestion, in nearly all cases the newcomers still have a better quality of life and range of opportunities than they had in the North East. Migration to the south is a trend that is likely to continue unless considerable investment is made in the north to improve life for those living there.

Differences or disparities in human welfare and quality of life occur within countries as well as between them. Unit 4.5 (page 109) looked at differences in development levels between two regions in Brazil – the North East and the South East – using a range of indicators. Source 1 here provides additional data of a human welfare nature.

South East Brazil is at the centre or **core** of the country's economy. It is easily accessible with a more equable climate than in the north. Most of Brazil's major cities are found here including São Paulo, Rio de Janeiro and Belo Horizonte – all major industrial centres. North East Brazil (together with the North region) is more remote and known as the **periphery**. It is an area prone to drought with a high proportion of the population working as farmers. Whilst the South East has attracted a great deal of investment especially from overseas, there has been very little in the North East. The Brazilian government has tried to spread development and wealth away from the South East. Regional development agencies have been set up (Source 1).

Like other LEDCs, Brazil is a land of enormous inequality and huge contrasts. The poorest 20 per cent of the population have 2 per cent of the country's wealth. There are many poor rural dwellers and poor shanty town dwellers with very low incomes and difficult quality of life. A quarter of Rio de Janeiro's population live in its shanty towns, called favelas. The country has rich urban minority. A few people have high incomes, many times greater than the country's poor. The richest 10 per cent of Brazilians have 51 per cent of the country's wealth.

Sheffield, UK

Sheffield is one of the UK's larger cities, with a population of around half a million people. It is a good example of a city with a strong socio-economic divide. Traditionally, it was a steel-making city, with steel factories being concentrated in the Lower Don Valley, but de-industrialisation hit the city in the 1980s. Some specialised steel is still made, but it is leisure, sport and other service industries, encouraged by Sheffield Urban Development Corporation in the 1980s and 1990s, that are helping to regenerate this area and the city.

Source 1 shows how infant mortality rates vary between the districts of the city. Source 2 takes two of these districts: Hallam in the west, which is one of the most well-off city suburbs in the UK, and Darnall in the east, a relatively deprived district.

More people are out of work in Darnall and fewer have well-paid, secure jobs and own a car. Far fewer young people in this district go on to university and so reduce their chances of improving their future lives. The lower life expectancy in Darnall, compared with Hallam, reflects the different lifestyles and environments in the two districts.

Most people living in Darnall do not own their home. Many of the houses there differ from the owner-occupied houses in prosperous Hallam – in size, style, age and facilities. In the larger, detached homes typical of much of Hallam, where many of the residents are university graduates whose children also generally go on to university, there are higher rates of car ownership and longer lives are being lived.

Ethnic minority groups tend to be concentrated in the poorer districts of UK cities. There is a large South Asian community in Darnall.

Darnall was home to many factory workers in Sheffield's old industrial days. The rich then lived in the clean air on the higher ground in the west end, before it was polluted by the factories in the east end.

Darnall is close to the Lower Don Valley, with its new businesses and improved transport and environment, and is now having millions of pounds of the UK government's Single Regeneration Budget spent in it. This money, and money from the City Council, is being used to build new houses, clear sites for businesses to set up and to tackle the area's social and environmental problems.

Source 1 Infant mortality rates in Sheffield, 2001

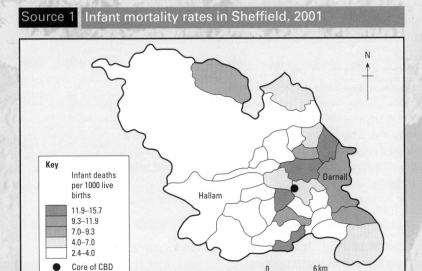

Key
Infant deaths per 1000 live births
- 11.9–15.7
- 9.3–11.9
- 7.0–9.3
- 4.0–7.0
- 2.4–4.0
- ● Core of CBD (city centre)

0   6 km

Source 2 Hallam and Darnall: a comparison

	Hallam	Darnall
Owner-occupier households	94%	27%
People unemployed	6%	19%
Households where the head is in a professional/ managerial job	54%	22%
Households without a car	20%	56%
18–19-year-olds entering university	62%	8%
Life expectancy	79	71

1 a Describe the three main methods used to measure quality of life.

  b What other factors, which cannot be measured easily, contribute to quality of life?

  c Quality of life varies considerably across the world. Describe some of the causes of these differences.

  d i What can be done to lessen inequality?

    ii How important is the role of free trade in this?

2 a i What do we mean by 'aid'?

    ii Why do many LEDCs need aid?

  b Name the four main types of aid and write a sentence to describe what each one means.

  c List the advantages and disadvantages of giving and receiving aid.

  d What is 'tied' aid? Why can it be a problem?

3 a What role does the World Bank play in trying to improve human welfare and quality of life?

  b Name some of the specific areas the World Bank is involved in.

  c What is an NGO?

  d i What do Oxfam see as being the right of every individual?

    ii How does Oxfam spend the money it raises?

4 a i Describe world population growth between 1000 and 2000 AD.

    ii Describe how it is projected to grow between 2000 and 2050 AD.

  b i What is meant by the earth's carrying capacity?

    ii Describe the different ways in which we need to use land to sustain and provide for people.

  c How and why is empowering women expected to affect future population growth?

5 a Write a brief report explaining China's population policies since 1970. Explain the consequences of these policies.

  b Describe the main differences in quality of life between North East and South East Brazil.

  c i Why do so many Brazilians migrate to the cities of the South East?

    ii How could this be stopped?

6 a Give the meaning of the term 'infant mortality rate'.

  b Describe the variations in infant mortality in Sheffield.

  c Suggest why:

    i people's income level can affect the infant mortality rate

    ii infant mortality rates affect people's quality of life.

  d Explain how schemes to redevelop inner-city areas like Darnall can benefit the lives of local residents.

Appendix A
Preparing and revising for your exams

Before you start
Before you can revise properly for your final written exam paper(s), you need to:

- ensure that all your classwork and homework has been completed correctly so that you have a good base to revise from – if you have work missing, check with your teacher
- know which level you have been entered for – Foundation or Higher
- be aware of whether you have one or two written exam papers. If you complete and submit a unit of coursework, you will only have one written exam; if you do not submit coursework, you will sit two written papers
- know the exact date, time, location and length of your written exams, what content is covered by each paper and what you can take into the exam with you, for example, calculator, ruler, etc.

All the above may seem quite obvious, but they can help you start to feel properly prepared to start your revision. You can then start to plan your revision. It is very likely that Geography will be just one of the subjects you have to revise for, so you need to look at all the exams you will be taking before you construct a revision plan or timetable. However, unless your exams are a spread out over a long time, it is advisable to allocate time every week for each subject.

Planning
Many people find it useful to plan a revision timetable over several weeks. If you do this try to:

- revise in blocks of 30–45 minutes
- choose times of day/evening which suit you best
- timetable regular breaks of 10–15 minutes between sessions, with proper, longer breaks for meals
- write out your timetable. Display a copy somewhere else where you live for everyone to see – so they know when to leave you in peace … or to remind you that you should be revising!
- find somewhere to revise where you will not be disturbed or interrupted easily.

Content
As well as planning your time – and breaks – you need to decide what parts of the course/topics you need to revise. Whatever level you are entered for, the first written exam paper covers the three main units, each containing two topics and the fourth unit containing three topics. These are:

- **Unit 1: People and the Natural Environment**
  1A Water
  1B Hazards
- **Unit 2: People and Work**
  2A Production
  2B Development
- **Unit 3: People and Places**
  3A Migration
  3B Urban Environments
- **Unit 4: Global Issues**
  4A Fragile Environments
  4B Globalisation
  4C Human Welfare

Both the Foundation and Higher level first exam have the same structure, although the Foundation exam is 1 hour 45 minutes and the Higher exam is 2 hours 30 minutes. Both papers have two sections:

- Section A: six **compulsory** short-answer questions, two each covering Units 1, 2 and 3
- Section B: a **choice** of one question from three, one each covering the topics in Unit 4.

If you have not submitted a coursework assignment, you will sit a second, 1 hour long exam paper. These questions test what you have learnt from the range of fieldwork and practical work undertaken as part of your IGCSE course. There are three compulsory questions including:

- two skills-based questions
- one enquiry-based question.

Whether you are sitting one or two exam papers, you will need to divide up your revision time so you cover the content of Units 1–4 for the first exam (Paper 1F or 2H), and the range of skills and content you carried out for the second exam (Paper 3). It is important that you cover/revise **all** the units, or there will be questions you may not be able to answer

Revising

It is unlikely that your IGCSE geography exam is the first exam you have ever sat. You should be able to learn from your previous experience of revising and sitting exams. What worked well for you? Were there things that did not work? Use and build on the methods that have been successful before – and learn from those which may have been unsuccessful. Remember, there are many ways to revise and different methods suit different people. You have to learn which are best for you – a range of tips are given below:

- **Glossaries/definitions** A good starting point for revising each topic are the glossary words. List these from each topic and write your own definitions, then check (and correct if necessary) from the book.
- **Case studies** For most if not all the topics you have studied you will have looked at actual examples of places in the world which help illustrate them, for example urbanisation in São Paulo, Brazil. It is important that you revise thoroughly such case studies as you will need to use them in your exam.
- **Crib cards** Many students find it useful to write key facts (or definitions) on separate file cards. You can use bullet points and mnemonics (rhymes or phrases) to help summarise

information. These crib cards and summaries can be used to test yourself on the content for each topic.

- **Mind maps/spider diagrams** These are **active methods** to help you revise. Rather than just reading through notes, start by writing a topic, for example 'earthquakes', in the centre of a sheet of paper. Then start to add key words or diagrams/symbols from the main word, such as 'causes', 'effects', 'case studies', etc. These diagrams are often easier for you to recall than lists of facts.
- **Practise geographical skills** Maps, diagrams and data are important geographical tools. You may be asked to measure, draw, complete or analyse information from all of these sources.
- **Skimming and scanning** When you read through your work, learn to read quickly to find the main points which you can then summarise as bullet points, crib cards or spider diagrams, making it easier to remember.
- **Past papers** Make sure you are as familiar as possible with the style of questions you will have to answer. Look at and practise by using past exam papers or specimen papers.
- **Review last session** Before you start each session of revision, spend 5 minutes reviewing the previous session to help reinforce what you have revised.
- **Active revision** Just reading through your work is not likely to help you remember as well as if you make sure your revision is **active**. Use the tips listed above to help you record/write down facts and information; test yourself or work with other people and test each other.

In the exam

- Check the rubric carefully. These are the instructions on the front of the exam paper which tell you which questions you must answer and whether you have a choice. Only attempt the correct number of questions. Where you have a choice, choose the question(s) you think you can do best on the topic(s) you have studied.

- Time – divide your time equally between the questions, allowing at least 5 minutes to read through them before starting your answer. You do not have to answer the questions in order – you may want to answer the questions you think you can do best first.
- Skills – geography exams test a range of skills, so be prepared (and equipped) to use maps, diagrams, data and diagrams.
- Read each question carefully, paying particular attention to command words which help you focus your answer. Words like 'complete' or 'describe' are basic commands, usually found in the first part of a question. Terms like 'explain' or 'give reasons for' are higher command words, are usually in later parts of questions and are often worth more marks.

And lastly – good luck. If you have completed the course and revised thoroughly, you may even enjoy your exam!

The coursework element is worth 20 per cent of the final mark. Not all centres/students undertake the coursework option: those that do not sit a second exam paper, common to both Higher and Foundation candidates and also worth 20 per cent. It is a practical paper with three compulsory questions (two skills-based and one enquiry-based), testing skills and experience gained from practical work and fieldwork undertaken during the course.

Coursework in Geography is based on fieldwork. This is work carried out first-hand by you outside the classroom, usually in your local area ('in the field'). It involves investigating a particular issue or topic, testing a hypothesis or question, by collecting and recording data and information using a variety of geographical skills and techniques to draw conclusions. You must collect a range of primary data to use in your coursework study, but can also refer to and use secondary data.

Structure and marking

Your teacher uses a clear, set mark scheme when assessing your finished work. To gain top marks, your coursework needs to have a clear structure. The marks are allocated as follows:

- **Introduction and aims (5 marks)**
  This is where you write about the purpose of your work, for example the question, issue or hypothesis you are going to investigate and where and how you are going to carry it out.
- **Data collection (15 marks)**
  These marks are given for how well you collect and record data. You should describe how you carried out your work, explaining and justifying why and how you did so; for example, how you decided what questions to ask and how you recorded the answers. You should also explain any problems you encountered, and the impact this may have had on your results.

- **Data presentation (15 marks)**
  Having collected the data, it is important that you present it clearly, neatly and accurately. You should include a range and variety of methods, relevant to the task. Tables, maps, etc. should be clearly titled, with keys and scales as required. To score high marks, you need to include more complex methods of geographical presentation, such as annotated sketch maps and/or photographs, overlay diagrams, composite graphs or scattergrams. However, these need to be appropriate to your study and help you present your findings more effectively.
- **Analysis and conclusions (15 marks)**
  Having described what you did and why you did it, this section of your work should set out to describe and explain what the data you collected shows. You should analyse your results to see to what extent they satisfy the problem or question you set out to investigate. You should identify the different data collected and how it may link together.

  Your conclusion should state to what extent your investigation and the methods you have used have answered the question you set out prove or disprove. You may also consider what else you could have done to improve the investigation – what limitations there were and whether these could be overcome or not.

- **Planning and organisation (10 marks)**
  These marks are awarded for how well structured and presented your final coursework is. It should be clearly and logically set out, with the clear purpose of the chosen investigation evident throughout. Data and other non-text should be cross-referenced in the text, pages should be clearly numbered and use made of title, contents and index pages, plus a bibliography where appropriate.

A limit of 2000 words has been suggested. As a rough guide this could be broken down as follows:

- Introduction and aims   300 words
- Method (data collection)   400 words
- Results (data presentation
  and analysis)   900 words
- Conclusion and evaluation   400 words.

Before you start

The first section sets out clearly how your coursework is marked, with a clear structure laid out for you to follow. This section aims to help you fully understand what is expected from you before you get started. The first thing you need to do is decide what your investigation will be about. It is quite likely that your teacher will choose a suitable topic which can be studied locally. This may cover a physical or human geography investigation, depending on where your school or college is based. You will probably work as a class or group to collect the data you need for your study. This is often helpful because you can collect a wider range of data than if you were working on your own or in a pair. Larger samples, when completing questionnaires or surveys for example, are likely to give more accurate results.

Remember:

- Your teacher will have carried out a thorough risk assessment of the area where your fieldwork/data collection will be take place. Even so, you should not work on your own, even in a 'safe' environment. It is safer to work in a small group. You must also be aware of other possible dangers specific to area you are working in, for example if you are studying a river and measuring flow, depth, etc.
- Although you may collect data as a group and share the information, when you write up your coursework, especially the analysis, conclusion and evaluation, it must be your own work.
- Your teacher can guide, encourage and advise you, especially in the methods of data collection, but he or she cannot help you

analyse your results and present your conclusion(s). Listen carefully to the help and advice your teacher gives and apply it where relevant to your work.
- Coursework is 20 per cent of your final mark/grade.
- There is not a word limit on your coursework, but the examination board recommend no more than 2000 words and they emphasise that quality is better than quantity.
- Up to 10 marks are awarded for presentation, so ensure your work is neat and tidy. If you can, use a computer to word process and present your final work.
- You need to show that you understand a variety of geographical skills and techniques and can use them appropriately. You will be expected to use a range of maps, diagrams, photographs, data and tables.
- If possible, use a range of ICT software and hardware to produce and enhance your work and integrate it into your study. This could include searching the internet for relevant information, selecting and using suitable maps and images, using specific software, such as databases or spreadsheets, to organise and present data, annotating maps and diagrams and using digital cameras to produce annotated photos.
- Your final piece of work must be presented on A4 paper. You should have a title page and include your centre and candidate numbers on the front cover. Present your work in a simple lightweight folder **not** in plastic wallets or ring binders.
- Edexcel have produced an investigation planning sheet (page 215) which can help you (and your teacher) to plan the structure of your study.

A sample fieldwork investigation

In any investigation testing a hypothesis or question, it is best to concentrate on one specific question based on a fairly small local area, such as a village or local river. The following sample fieldwork investigation can be used as a guide to your own study.

Edexcel's investigation planning sheet

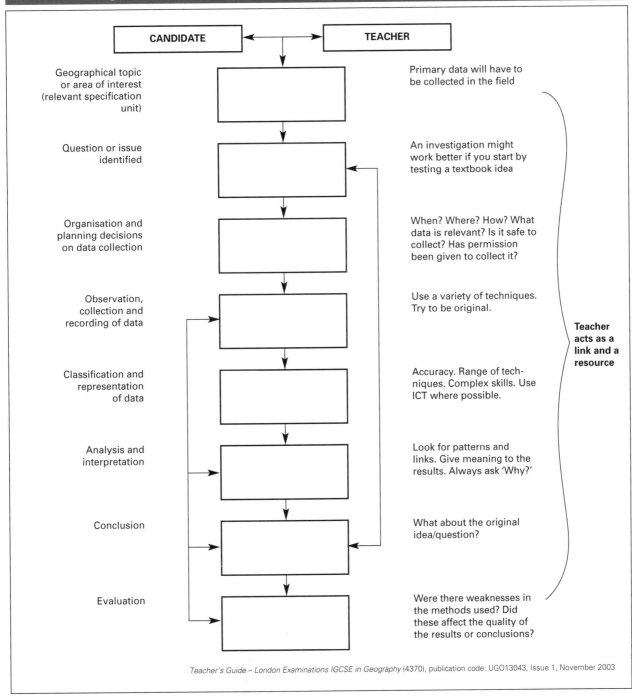

Teacher's Guide – *London Examinations IGCSE in Geography* (4370), publication code: UGO13043, Issue 1, November 2003

Question/issue:

The visual impact of a new wind farm

Aims:

- *To determine the views of local residents*
- *To discover whether those views differ as distance from the wind farm increases*
- *To assess the overall impact of the wind farm*

For the purpose of this sample fieldwork investigation, it will be assumed that there is a new wind farm located within 5–10 km of your school/college which has been operating for two years.

Planning and preliminary work

Before you plan how to carry out your investigation, think carefully about the issue –

what you already know about it and what you need to know before you begin. It is important to know and understand about the issue generally, for example energy resources, as well as the local wind farm itself.

This investigation is based on one of the topics dealt with in Unit 2A, People and Work – Production, so this is a good place to start. Read the parts of the unit which deal with energy resources. Look at the differences between renewable and non-renewable energy sources, and the case study about wind farms. This should give you an excellent introduction and overview to the topic. Think about the advantages and disadvantages of different energy sources, the impact each has on the environment and how important they are in the overall production of energy.

You will probably also know something about the local wind farm itself. You and others in the area are likely to have your own opinions about it, good and bad. Both facts and opinions are important in this study, as much of the investigation is based on what people think about the wind farm. However, before you consider local opinion, you need to know as much factual information about the wind farm.

Collecting data and evidence

You have two types of information to collect – factual (objective) and opinion (subjective).

Factual

What do you need to know about the wind farm? Maps are a good starting point. Your study needs to show where it is located within the area, for example in relation to the physical landscape, settlement and transport. A detailed map of the site itself is also important, showing the individual turbines, distance apart, links between the turbines and generator and on to the electricity grid and/or customers.

Who built and owns the wind farm? How much energy does it produce daily, weekly, yearly? What proportion of the energy needed locally is supplied by it? Who uses the energy? Why was this site chosen to construct it on? How many people work here? What does it look like – how many individual turbines are they and how high are they? How is the wind farm accessed, for example did new roads have to be built?

Much of the factual information can be found from secondary sources, such as local newspapers, local council (who will have given the company permission to build the wind farm), the internet or from literature produced by the company or pressure groups. Libraries are an excellent source of information, especially about local issues. Be careful when collecting factual information that it is not based on opinion.

You are advised to make a variety of sketches and photographs. These should help your investigation, so should be of the actual wind a farm and various residential locations at different distances from it.

Opinion

A major part of this investigation is to find out what local residents think about the wind farm in terms of how it looks. This is very subjective. To collect this information, you will have to construct one or more questionnaires and get as many local residents as possible to complete them. You may also want to interview a selection of key

Please indicate how you feel about the wind farm by ringing the relevant number.					
Attractive	1	2	3	4	ugly
Quiet	1	2	3	4	noisy
Essential	1	2	3	4	not essential
Good for the area	1	2	3	4	bad for the area
Too close	1	2	3	4	far enough away
Better than alternatives	1	2	3	4	worse than alternatives

1 and 4 = feel very strongly
2 and 3 = feel less strongly

people in more detail, for example the wind farm manager, someone who works there, a local councillor/politician, planner and leader of residents or pressure group. Start with your basic questionnaire then add an extra section of questions.

Start by making a list of the precise information you need to know from the person completing the questionnaire. For example:

- *Where do you live?*
  You need to know the exact location so you can know how far away they live from the wind farm. One of the aims of your study is to see if views about its impact vary according to distance from it.
- *What do you think of the wind farm?*
  Be careful not to ask this question in this form, i.e. an 'open' question, as it will make collating the responses and presenting them very difficult. Instead give a choice of answers, such as 'very unhappy', 'unhappy', 'happy', 'very happy'. Try to have an **even** number of responses so people can't consistently sit on the fence by giving the 'middle view', for example 'satisfactory'/'don't mind', etc. It makes people really think whether they agree or not!

Alternatively use a bipolar scale like this:

Bipolar scales are usually quite quick to complete, cover a range of opinions and the responses are easy to collate.

Do leave room on any questionnaire for other or different responses – some answers here may add considerably to your investigation, especially if the interviewee has strong opinions or connections to the wind farm, for example uses the energy/works there. However, you will need to think carefully about how you present additional comments and what weight you give to them.

Remember when you ask people to answer questions, you should always be polite, work in a pair (or in sight of another member of your group) and tell the interviewee why you are

carrying out the work. Do not collect names or addresses – the street name is sufficient for location (unless it is a very long road!). If you think the age of the person is relevant, do not actually ask them but try to decide what age group they belong to.

Although it takes more time, it is better to go through the questionnaire personally with local residents, rather than leave it with them or post them through doors. It is not usually a good idea to call door to door, unless you know the person living there. Instead, choose suitable locations where many local people go, for example the main shopping area. You should seek and take advice from your teacher about this.

Read the aims of your investigation again. Have you made sure that you can collect all the information you need from your questionnaires? Do you need to collect other data in different forms?

Presenting data and evidence

Once you have collected all the data you need, you have to decide how best to collate and present it. Again, look at the aims of your investigation. Decide which data fits which aim best. It may be a good idea to present this separately for each aim, so it is clear what you are trying to show. If you have collected a lot of responses as a group, using a simple database package on a computer is a good way of collating it. Bear this in mind **before** you construct your questionnaires, so the responses can easily be entered.

You must present your data in a variety of ways. For example, you can use a map to show the how far each your interviewees live from the wind farm (aim 2). You will then need to decide how you can then use this map to show differing views about it. You can also mark on this map, or second one of the same area, the locations of any photos or sketches you include.

Graphs can be hand or computer drawn – preferably a mixture to show your range of skills. Try to choose a type of graph which presents your data clearly, for example bar, line, pie chart, scattergram, etc. but one which is appropriate.

Make sure all data is clearly titled and referred to in your writing. Adding a number is the best way of doing this, for example Graph 3, Map 2, etc. This is very useful if you refer to the same data more than once in different section of your work, especially in the analysis and conclusion.

Always include the original data you collected – ask your teacher for advice here, but this is often better presented as a section at the end of your final study, although you should refer to it in your writing.

Analysing data and evidence

This is the section where you pull together the data you have presented to show to what extent it answers the aims of your investigation. Again, refer to the aims in detail and make sure you describe and explain clearly what evidence your investigation has collected and what it shows. What is the range of views about the visual impact of the wind farm? How did you assess this? Are the people who feel most strongly (negatively) those who live closest to it? How do their views compare with those living further away? Overall what is the impact?

When you are analysing your evidence, refer to relevant evidence clearly, for example 'Map 2 (p8) shows that those living nearest the wind farm feel most negatively about it.' Your teacher cannot advise you on this section – take care to write clearly and in a structured way so you can gain as many marks as possible (see Structure and marking, page 213).

Writing your conclusion and evaluation

This should be fairly short and concise. It should focus on the main issue – the visual impact of the wind farm – and summarise your findings. To what extent does the wind farm impact on the environment/look of the area? Is there a link between what people feel and how far away they live? From your analysis, what conclusions have you come to about the overall impact of the wind farm – good and bad?

Do not forget to include a few comments evaluating your work. Would you have liked to have had access to other information but could not? What was this? How would it have improved your study? Were there any major difficulties?

Glossary

abrasion process of river erosion in which pebbles are used to break off pieces of rock from the bed and banks of the river

acid rain a cocktail of chemicals (e.g. sulphur and nitrogen oxide)

agribusiness large-scale intensive commercial farms

aid loans and goods given to the LEDCs by the MEDCs

attrition process of river erosion in which particles are reduced in size as they are being transported

balance of trade the value of exports minus the value of imports; there may be a trade surplus or a trade deficit

bilateral aid aid given by one government directly to another

bio-degradable applies to a material that can be decomposed by the action of living organisms

biodiversity the number and range of plant and animal species in a specific place

biogas a methane-type gas produced by fermenting animal dung, which can be used as a source of energy

biomass the total amount of organic matter in an ecosystem

biome a global ecosystem, for example savanna or tropical rainforest

birth rate the number of live births per 1000 people per year

brownfield sites land prereously built on, often old industrial land or land once used for housing, now available for redevelopment

carrying capacity the total number of people that can be supported by existing natural resources

CBD Central Business District or city centre

CFCs chlorofluorocarbons, a greenhouse gas

commuter someone who travels daily into and back from work, often from the suburbs to the CBD

condensation water vapour in the air is converted into water droplets by cooling

constructive plate margin where two plates move away from each other

corrosion process of river erosion in which rocks are worn away by chemical action

counterurbanisation the movement of people away from urban areas

death rate the number of deaths per 1000 people per year

deforestation the removal of forest by burning or cutting

demographic to do with population change

demographic transition model diagram which shows the relationship between birth rates and death rates

demography the study of population

density number of people per square kilometre

deprivation a measure of how poor or badly off people are, putting them at a disadvantage in relation to others

destructive plate margin where two plates collide with each other

development agency an organisation set up by the government to organise and promote development in a region

development indicators a range of data used to compare levels of development, for example life expectancy, birth and death rates

discharge the volume of water in a river at any given time

distribution (of a population) where people are found and where they are not found

drainage basin the area of land drained by a river and its tributaries

ecotourism tourism designed to have a minimal impact on an environment, helping conservation

edge city a new purpose-built city constructed on the edge or periphery of an existing city

emergency aid aid given after a natural disaster such as a flood or earthquake

ethnic cleansing the mass killing or eviction of a specific religious or ethnic group by a different group

evaporation when water droplets are heated up to be converted into water vapour

evapo-transpiration when water droplets from trees and vegetation are converted by heat into water vapour in the air

exports goods sold abroad

famine a shortage of food causing malnutrition and hunger

fault fracture in the earth's crust

focus underground source of an earthquake

fossil fuels coal, oil and natural gas

genetically modified crops crops whose make-up has been altered genetically to increase their yield and resistance to diseases and pests

global warming an increase in the average temperature of the world's atmosphere

green belt areas of countryside around a city in which most new types of building are forbidden

greenfield location area of land that has not been developed for industry

Gross Domestic Product (GDP) the total value of goods and services produced by a country

Human Development Index (HDI) a comparative measure of quality of life based on a range of development indicators

Human Poverty Index (HPI) a set of data (human development indicators) used to assess relative poverty levels

hurricane a severe tropical storm with winds over 120 km/hour and very heavy rain

hydraulic action process of river erosion in which the force of the water breaks off rock

imports goods bought from abroad

infiltration water seeping vertically downwards through the soil

infrastructure the basic framework of faciities that service an industrial country, for example roads, electricity supply

intrusive volcanic activity magma solidifies beneath the surface of the earth

irrigation the artificial watering of the land

lag time the difference in time between peak rainfall and peak discharge

lava molten rock on the earth's surface

magma molten rock beneath the earth's surface

mass movement the movement downhill of soil and rock

megacity a city with a population of over 10 million

migration movement of people

morphology layout and features which can be seen

multilateral aid money given by several governments to international institutions who pass it on to the LEDCs

natural increase an increase in population caused by the birth rate exceeding the death rate

non-biodegradable material that cannot be decomposed

non-government organisations (NGOs) groups, often charities, not supported by government funding, supplying help, advice and aid to those in need

ozone layer layer of the atmosphere which protects the earth from ultra violet radiation which can cause skin cancer and destroy organisms

padi field a flooded field where rice is grown

Physical Quality of Life Index (PQLI) three indicators – literacy, infant mortality and life expectancy – used together to compare relative wellbeing

plantation a large estate where one main cash crop is grown, often run by a transnational corporation

precipitation water from the atmosphere in any form; rain and snow are examples

quaternary industries these include industries that provide specialist information and expertise

refugees people forced to move from where they live to another area

river regime the pattern of variations in discharge in a river over a year, often shown as a graph

runoff water that flows overland across the surface into and in rivers

rural-urban fringe area around the edge of a town or city where the built-up area ends and the countryside begins

set-aside a European Union farming policy whereby farmers are paid to leave some of their land fallow

Single Regeneration Budget funds provided by the UK government to local councils to redevelop run-down areas

soil erosion the wearing away or loss of soil mainly due to the action of wind, rain or running water

structure (of a population) the relative percentages of people of different age groups, usually shown on a population pyramid

subsidies financial support to assist an industry or business to remain competitive

tariffs customs duties charged on imported goods

tectonic plates large segments of the earth's crust which move across the globe

trade deficit where a country imports more goods than it exports

trading blocs groups of countries who join together for trading purposes

transnational corporations (TNCs) large companies which have branch plants throughout the world; their headquarters are often found in more economically developed countries

UNHCR United Nations High commission for refugees – international body which works for and helps refugees

Urban Development Corporation an agency set up by the UK government in the 1980s to regenerate deprived areas like inner cities

urban renewal the rebuilding or renovation of old, run-down urban areas

urbanisation growth of towns and cities leading to an increasing proportion of a country's population living there

Index